An INTRODUCTION to MOLECULAR BIOLOGY

Robert C. Tait, Ph.D.

Department of Neurology, University of California at Davis, 1515 Newton Court, Davis, California, 95616 USA

Horizon Scientific Press
P.O. Box 1
Wymondham
Norfolk NR18 0EH
England
http://apollo.co.uk/a/horizon

British Library Cataloguing-in-Publication Data

A catalogue record for this book is available from the British Library

ISBN: 1-898486-08-5 (paperback)
ISBN: 1-898486-09-3 (hardback)

Printed in England by Ipswich Book Company, Suffolk.

Books of Related Interest

Current Innovations in Molecular Biology

Volume	Title	Date
1.	**Molecular Biology: Current Innovations and Future Trends. Part 1.** A.M. Griffin and H.G. Griffin (Eds.)	1995
2.	**Molecular Biology: Current Innovations and Future Trends. Part 2.** A.M. Griffin and H.G. Griffin (Eds.)	1995
3.	**Internet for the Molecular Biologist.** S.R. Swindell, R.R. Miller, G.S.A. Myers (Eds.)	1996
4.	**Gene Cloning and Analysis: Current Innovations** Brian C. Schaefer (Ed.)	1997
5.	**Genetic Engineering with PCR** Robert M. Horton and Robert C. Tait (Eds.)	1998

Distributors

U.K. and Europe
Horizon Scientific Press
P.O. Box 1
Wymondham
Norfolk
NR18 0EH
U.K.
Tel/Fax: 01953-603068
International: +44-1953-603068
Email: horizon@usa.net
http://apollo.co.uk/a/horizon

USA/North America
Horizon Scientific Press
c/o ISBS
5804 NE Hassalo Street
Portland
Oregon 97213-3644
Tel: (503) 287-3093 or
(800) 944-6190
Fax: (503) 280-8832
Email: orders@isbs.com

Australia/New Zealand
DA Direct
648 Whitehorse Road
Mitcham
3132 Australia
Tel. (03) 92107777
Fax. (03) 92107788
http://www.dadirect.com.au

Rest of World
Horizon Scientific Press
P.O. Box 1
Wymondham
Norfolk
NR18 0EH
U.K.
Tel/Fax: 01953-603068
International: +44-1953-603068
Email: horizon@usa.net
http://apollo.co.uk/a/horizon

Preface

In the 1970's, researchers in a number of laboratories collectively developed methods that allow the purification, the specific dissection, and the re-arrangement of deoxyribonucleic acid, or DNA, the chemical compound that stores genetic information in most biological systems. While these methods, which have come to be known as "recombinant DNA technology", were developed for use in molecular biology, the analysis of the structure and function of genes, they also allowed the construction of new combinations of genes that do not normally exist in nature. For example, part of gene A can be isolated and fused to gene B to create a new gene AB that combines parts of the function of each of the two original genes. These methods also allowed a gene to be purified from one organism and transferred to a second recipient organism, introducing new physical charateristics not previously present in the recipient.

These techniques have initiated a scientific revolution in the field of molecular biology and have led to tremendous advances in understanding both the structure of genes and how genes are regulated in the cell. The application of recombinant DNA methodology to medical problems has begun to yield benefits in the diagnosis and treatment of a variety of diseases. Current results also suggest potential benefits in the diagnosis and management of a variety of agricultural diseases and in the manipulation of the characteristics of various plant and animal species.

Perhaps because of the revolutionary nature of this approach to the analysis and manipulation of genetic information, the application of this technology has not been without controversy. When scientists initially working with these techniques perceived a potential biological hazard in the manipulation of genes in this manner, national guidelines were established to regulate the conditions under which experiments were allowed. Work that was perceived to be of potentially greater hazard, such as working with the genes of organisms known to cause human disease, was performed under more contained circumstances than other experiments. With the increasing amount of experience, these guidelines have been periodically modified to take into account the accumulated information regarding various types of experiment.

In recent years, public attention has been drawn to several areas of research in which recombinant technology has been instrumental in the development of products, such as the use of synthetically produced interferon in disease treatment, or the use of an "ice-minus" strain of bacteria in the prevention of frost damage on certain agricultural crops. In addition, there have periodically been allegations that recombinant DNA technology opposes the will of God, is a sexist tool, is a racist tool that will be used in the construction of a superior race, has been developed by foreign powers as a secret weapon for construction of biological warfare weapons, and a variety of other alarmist claims.

Scientists have long been noted for their use of terms that are specialized to a particular field of research. While the use of this terminology can be beneficial to researchers, at times it can appear to a novice that this jargon is designed to prevent the public from actually understanding what is going on in the research. Molecular biologists are particularly adept at the development and dissemination of new terms, a trait that certainly has not helped the public understand recombinant DNA technology and the field of molecular biology. When confronted with news items concerning this technology, it is therefore not surprising that a layperson can have difficulty assessing the validity of various claims.

One way to combat the lack of public understanding of this technology is to introduce gene manipulation concepts at the level of high school and community college biology. This text was developed in association with the "Short Course in Recombinant DNA Technology" that was offered to high school and community college science instructors over a period of seven years at the University of California, Davis. This course was a one-week lecture/ laboratory course consisting of a series of lectures that deal with the basic principles that constitute recombinant DNA methods, followed by simple exercises that demonstrate the purification of DNA and the construction and analysis of recombinant DNA molecules. The lectures were intended to help explain the terminology and simplify the jargon, and the methods used in the laboratory portion of the course were identical to those used by many scientists engaged in molecular biology research. In addition, a series of seminars and demonstrations by research faculty members and graduate students from UC Davis allowed participants to see how molecular biology research applies these methods to specific research problems and to interact directly with scientists active in molecular biology research. The overall course was designed to de-mystify recombinant DNA and molecular biology methodology and help participants in the course integrate the information into their own science teaching programs. Much of the information presented in this book was derived from the years of use and improvement of the information content of this continuing education course.

Some of the high school and community college instructors that have participated in these programs have been able to return to their own schools and effectively utilize the new information in their own teaching programs. This text is an attempt to provide a short, introductory course to the principles of recombinant DNA methods and their applications to molecular biology, presented in a format that facilitates the direct incorporation of blocks of information into existing teaching programs. Key concepts are summarized at the end of each chapter to facilitate design of lectures. A series of exercises that can be performed as either demonstrations or as class laboratory exercises, complete with detailed instructions and ordering information, is provided to help re-inforce key principles of recombinant DNA methods.

This text has also had significant input and undergone modification as a result of the years of teaching recombinant DNA and molecular biology methods to the graduate students, staff, and non-molecular biology faculty at UC Davis. The twenty years of experience of Dr. Tait as an acknowledged molecular biology "guru" at UC Davis and extensive collaboration with faculty involved in a wide variety of molecular projects have contributed greatly to the broad scope of this manual.

As is the case with any scientific method, recombinant DNA technology should not be considered as inherently "right" or "wrong", but should be considered as merely an investigative method that can be applied to a wide variety of biological problems. Accurate public assessment of the potential risks or benefits of a particular scientific approach requires understanding the basic methods under question. It is the hope of the author that this text will be of help to students in various fields of science and will also help improve public understanding of recombinant DNA methods and will help facilitate informed public assessment of the merits of the use of these methods in both basic and applied biotechnology research.

Robert C. Tait
University of California

Contents

1

Understanding the Scientific Method

"Let's start at the very beginning, a very good place to start." — Julie Andrews as Maria in *The Sound of Music*

The historical attitude of the non-scientist towards understanding science might be accurately depicted by the phrase "I'm not a scientist, I'm not going to be a scientist, they're going to do it anyway, so why should I care?" Good question. Why should you care?

The typical biological research scientist can be reasonably described as being concerned primarily with the question "Why does it work like that?" Achieving an understanding of the mechanisms by which cells function has for years been a primary goal of biological research. Research can be described as years of very tedious investigation, punctuated by moments of brilliant insight, followed by more years of tedious investigation, all driven by a compulsion to understand the solution to a problem.

In our technologically oriented society, scientific research is often viewed as a source of new information that can help provide solutions to problems, innovations to improve the quality of life, or new and improved commercial products. The scientific knowledge that develops as a result of biological research is by itself neither good nor bad, it is merely information. The uses to which this information is applied, however, can clearly be either helpful or harmful to society in general. It is the application of knowledge rather than the knowledge itself that should be of concern to all members of society.

New technology can clearly have long-lasting effects that are not readily apparent during the initial development and application of the methods. The internal combustion engine, for example, utilized a new hydrocarbon energy resource to mechanize and industrialize world economy. The generation of hydrocarbon smog and the ultimate dependence of world economies on the availability of oil, however, were complications of the new technology that were not readily forseen. In a similar manner, the development of atomic energy was hailed as a source of cheap, clean, readily available electric power that would revolutionize the world. In the wake of the nuclear accidents that occurred at Cheronobyl and Three Mile Island and with the accumulation of intensely radioactive wastes that cannot be easily stored or detoxified, many people now question the concept of nuclear power as a safe, viable energy source. Even a technological development as trivial as the fluorocarbon propellants used in aerosol cans has been accompanied by a growing concern that the propellants may be causing significant damage to the ozone layer that protects the earth from intense ultraviolet radiation. The list of real or potential problems that have arisen as a direct result of technology is long and continues to grow.

This does not mean that all technology should simply be abandoned for fear of potentially harmful future effects. The potential harm associated with the application of new technologies must be compared with the potential benefits of the methods. It is therefore appropriate for the general public to become sufficiently scientifically educated to be able to evaluate the potential risks and benefits of new technologies. The growing concern for the quality of the environment is an excellent example of how public education and awareness have led to public assessment and re-evaluation of long-established environmental policies. The incidents at Love Canal and many other toxics disposal sites have resulted in the modification of traditional waste disposal techniques, for example. Public awareness and understanding of the problems under consideration, however, are fundamental to accurate evaluation of potential problems associated with scientific technology.

The technology that has developed in the field of molecular biology has done far more than provide tremendous information about the workings of the cell. These methods have been rapidly assimilated by the industrial sector and applied to a wide variety of problems. Hormones and proteins are currently being manufactured by microbes grown in vats, viruses have been engineered to allow the production of vaccines, DNA detection methods have been used to detect disease-causing organisms, and DNA "fingerprints" have been used in court to establish guilt or innocence of suspects. The technology is so powerful and has such widespread application that it will not simply go away if ignored long enough. Since the methods are likely to be a long-lasting, fundamental aspect of biological research, biomedical technology, and bioindustry, it seems appropriate to obtain a basic understanding of the principles and methods that make up this extremely powerful technology.

To understand why recombinant DNA methods and molecular biology have risen to such a level of scientific significance, it is helpful to first examine the basic process by which scientific research is conducted. The human race has for centuries tried to understand the principles that direct the functioning of the world and has developed different approaches for explaining these principles. Explanations that account for natural phenomena might be thought of as falling into one of three general categories: myth, religion, and scientific explanation.

Myth, religion, and science

While myth, religion, and science are similar in that all may involve potential explanations for natural phenomena, they are fundamentally different in the mechanisms by which these explanations are generated (Figure 1.1).

A myth is generally considered to be a notion or story that is widely accepted, but may or may not actually be true. Virtually every major culture has an origin myth, for example, that describes the beginnings of the world as perceived by that particular culture. In some origin myths,

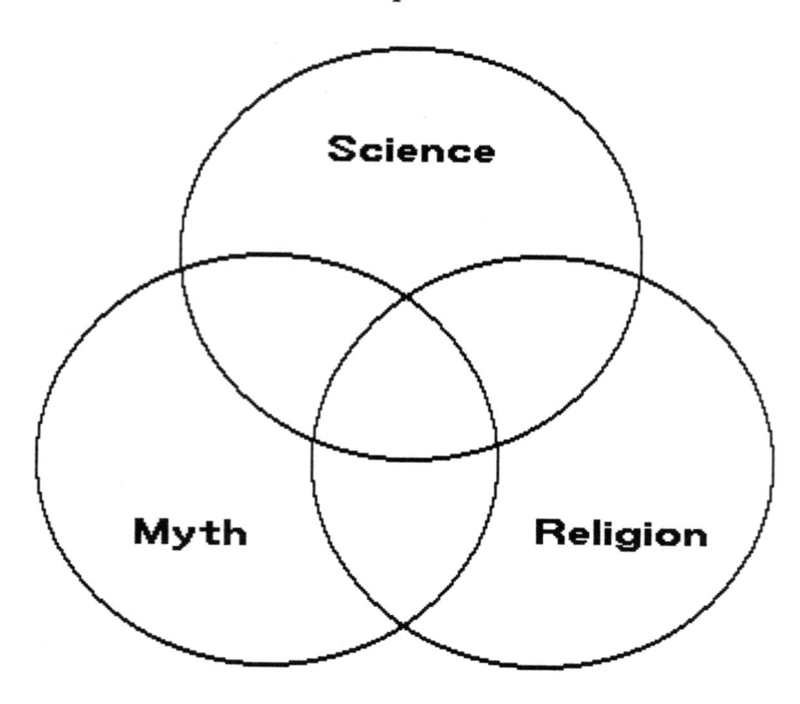

Figure 1.1. The relationship between myth, religion, and science in the attempt to explain an observed phenomenon. Note that failure to adhere to the scientific method can cause "scientific" explanations to overlap with either myth or religious explanations.

a supreme being forms the world and all in it from the darkness, in others from water or mud. The details of these stories are rarely in agreement, suggesting that not all of the stories can possibly be factually correct; these stories share in common the attempt to explain how the world came to be. Many myths are widely recognized to actually be untrue. Walking under a ladder, a black cat crossing your path, spilling of salt, and breaking a mirror are all occurrences that, according to myth, will bring bad luck. While some people adhere to the belief that these incidents forebode bad luck, many other people pay no attention to these myths.

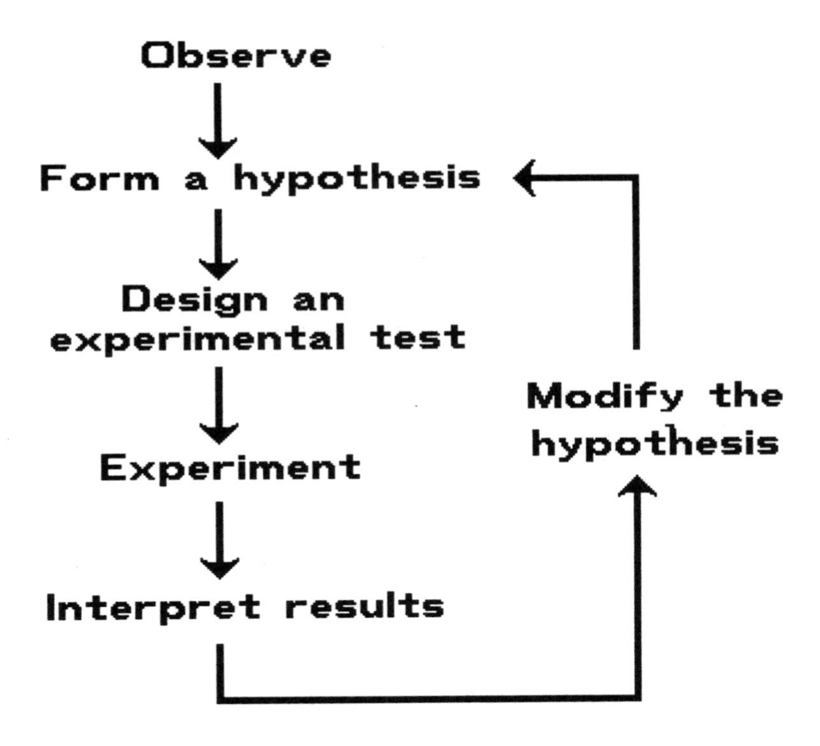

Figure 1.2. The scientific method as cyclic process of evaluation and testing of hypotheses.

Information that is considered to have been transmitted by divine intervention, the direct inspiration of an individual by a supreme being, often takes the form of a religion. While many people may consider religion to be a form of myth, religion is somewhat different in that the principles presented in the religion are generally intended to be accepted as true as a matter of faith. While it may contain explanations for the functioning or existence of the world, a religion generally also contains rules of social, moral, or ethical conduct that seem designed to help a culture interact with the world around it. As is the case with myths, most of the cultures in the world have developed religions. Because the very nature of religion usually demands acceptance without testing or questioning, differences in religious beliefs have probably been one of the greatest historical causes of major cultural conflicts.

It has been said that science has become the new religion of the twentieth century. While there may be an element of truth to this statement, scientific explanations of phenomena are based on what has long been called "the scientific method", a process intended to distinguish scientific explanations from both myth and religion. The scientific method is based on the formation and testing of a hypothesis, a potential explanation for an observation (Figure 1.2). Scientific research can be described as a cyclic series of steps performed in the order: observe, form a hypothesis, test the hypothesis, interpret the test results, and modify the hypothesis. Each time a test is completed and the results are examined, interpretation of the results often reveals aspects of a hypothesis that are not fully consistent with the experimental results. The researcher therefore modifies the original hypothesis to try and account for the experimental observations, designs a test of the modified hypothesis, and repeats the cycle. When all tests that can be designed produce results that are consistent with the hypothesis, that hypothesis is considered to be the best explanation for the original observation. With time, the hypothesis might come to be accepted as a "scientific fact". Note that the "test, interpret, modify" aspect of the scientific method distinguishes from both myth and religion any explanations that are derived by this method.

It is important to emphasize that the scientific method, when carefully applied to the analysis of an observation, does not generally **prove** a hypothesis, but merely generates a potential explanation that is **consistent** with all the tests that have been applied. The validity of a scientific explanation is limited by the number of tests that have been performed on the hypothesis. A large number of tests that support a hypothesis can lend confidence that the hypothesis is correct, but cannot prove that the hypothesis is actually true. It is equally important to emphasize that each test is designed within a framework of other observations and hypotheses that may have already come to be accepted as true. A single flawed test or assumption used to interpret a test can affect the validity of the scientific method, thereby invalidating the test of a hypothesis.

A critic of the scientific method could justifiably point out that, since tests are designed using scientific principles that have sometimes never been actually proven, the whole, interrelated network of scientific explanations regarding the functioning of the world is like a chain with one or more weak links. A complex scientific explanation for a natural phenomenon is only as valid as the least well-studied assumption used to interpret the results of scientific studies. Since many of the fundamental scientific "facts" used to design and interpret tests might be considered by some people to actually be scientific myths - widely accepted but essentially unproven beliefs or hypotheses to which the correct tests have not yet been applied - the validity of all scientific explanations should be seriously questioned.

It is precisely this criticism that drives scientific researchers to constantly test and re-evaluate hypotheses and to develop new methods for performing tests of widely accepted notions. The constant re-evaluation of hypotheses and assumptions and the interchange of ideas among different branches of scientific research are among the forces most responsible for the advent of new scientific technologies. The ongoing re-evaluation of ideas and the attempt to generate explanations that are consistent with all possible experimental tests, the key features of the scientific method, are what distinguish scientific explanations from both myth and religion.

Genetics and biochemistry set the stage for the advent of DNA technology

One of the most intriguing observations in biology is that a single fertilized egg cell is capable of giving rise to a complex organism possessing many different tissue types, each capable of performing a different subset of the processes necessary for the survival of the organism. The mature adult organism, in turn, is then capable of generating gametes, or sex cells, that produce progeny much like the

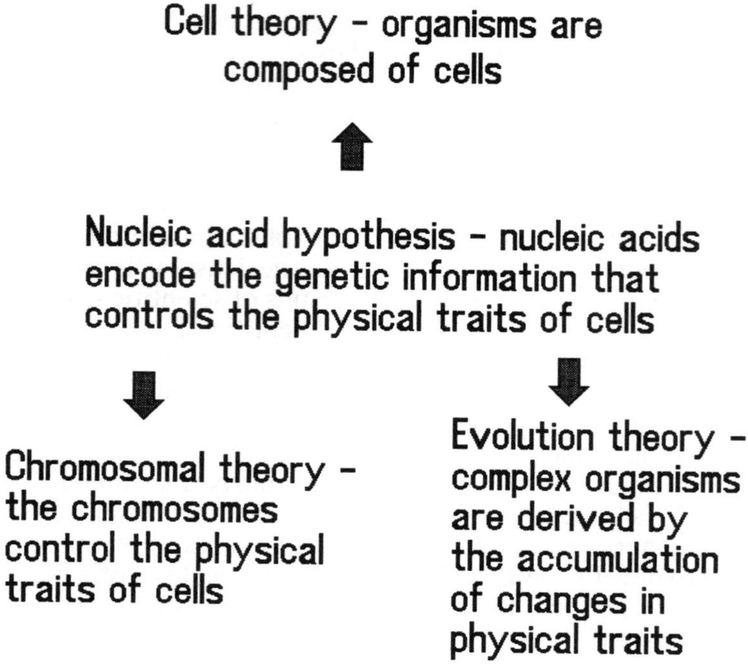

Figure 1.3. The nucleic acid hypothesis as an explanation for the cell theory, the chromosomal theory of heredity, and the theory of evolution by natural selection.

parents. The overall fidelity of this process is so high that the appearance of developmental abnormalities, such as a two-headed cow or a frog with six legs, is an occasion often accompanied by much publicity.

By the 1960's, the application of the scientific method to the study of genetics, the transfer of traits from parents to progeny, had given rise to three important hypotheses, or theories, that attempted to explain the heritability of traits: 1) the **cell theory**, suggesting that all organisms are composed of many individual cells, 2) the **chromosomal theory of heredity**, suggesting that the chromosomes within the individual cells control the physical traits of the cells, and 3) the **theory of evolution by natural selection**, suggesting that complex organisms are derived from more primitive organisms by a process of accumulation of changes in physical traits of cells.

It is important to always remember that these three fundamental biological concepts are not facts or laws, but are merely theories that must be continually tested to determine the limits of their accuracy. By the 1970's, the results of many independent studies performed by different researchers were all basically consistent with these theories, which could be unified in a composite **nucleic acid theory**: the physical traits characteristic of the cells in an organism are encoded by the nucleic acids present in the chromosomes of each cell and change in the structure and function of nucleic acid resulted in changes in cellular characteristics (Figure 1.3).

Further testing of the role of nucleic acids in the control of the physical characteristics of cells could be most rigorously accomplished only if the nucleic acids corresponding to specific genes could be isolated, characterized, and specifically modified to test the relationship between nucleic acid structure and gene function.

Summary

1. **Scientific explanations of observations are based on the scientific method, a process in which a hypothesis is proposed, experimentally tested, and modified to be consistent with the experimental observations.**

2. **Scientific explanations are no more valid than the tests to which they have been subjected.**

3. **The hypothesis that the structure and function of nucleic acids controlled the physical characteristics of cells helped drive the development of the recombinant DNA methods that enabled the isolation and analysis of nucleic acids.**

2

DNA and RNA as Genetic Material

"I yam what I yam and that's all what I yam." — Popeye the Sailor

That statement by the world's most well-known consumer of spinach summarizes one of the fundamental concerns of molecular biology research - why do cells function the way they do? Cells must possess several fundamental properties in order to be considered "living" - they must be capable of catabolism, or breaking down complex compounds to derive energy, of anabolism, or the synthesis of complex compounds from simple substrates, and of reproduction, or the production of progeny with traits similar to those of the parental cells. The cells must be able to sustain the variety of chemical processes necessary for the maintenance or survival of the cell itself and for the ultimate production of offspring.

By 1970, biochemical research into the chemical processes that take place in living cells had contributed greatly to the understanding of the cell theory that had been formulated based on the results of genetic experiments. This chapter provides a description of the basic hypothesis of cell structure and function and how the development of recombinant DNA technology was vital to the testing and refining of these basic notions.

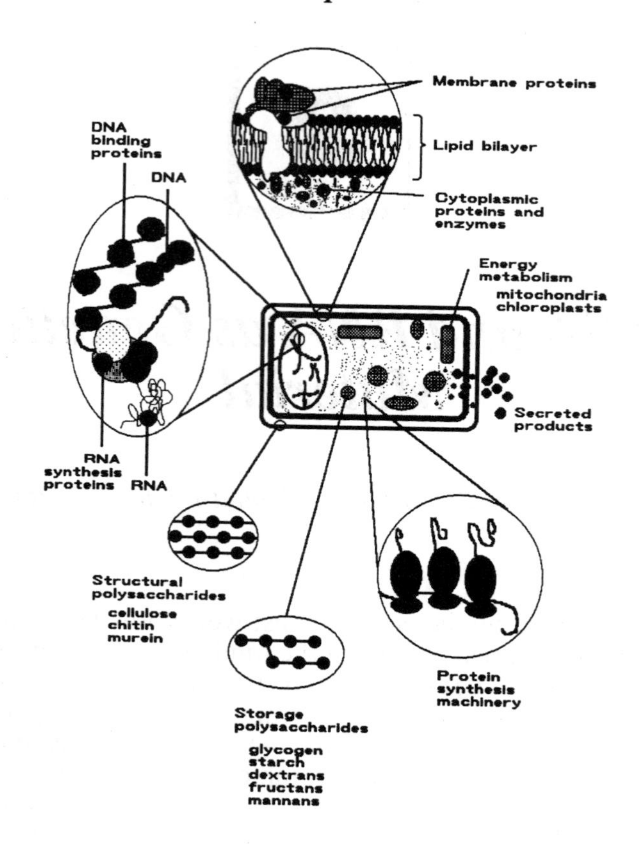

Figure 2.1. Genetically determined aspects of cellular metabolism. All biological processes are either directly or indirectly determined genetically. Genes are present on deoxyribonucleic acid (DNA) present in chromosomes that are composed of DNA associated with many specific DNA binding proteins. Ribonucleic acid (RNA) is synthesized directly from genes by specialized RNA synthesis protein complexes. Proteins are assembled by a complex of RNA and proteins and are converted to their active forms by other proteins. The membranes that surround and form compartments within a cell are double layers of lipids, or oils, that contain proteins. While the membrane proteins are genetically determined, the lipid components of a cell are not directly synthesized from genes, but are assembled by enzymes that are genetically determined. Although polysaccharides are not directly synthesized from genes, they are also assembled by genetically determined enzymes. The energy metabolism is also determined genetically, with some components of organelles like mitochondria determined by chromosomal genes and others determined by specific mitochondrial genes.

Table 2.1. Abbreviations for the twenty common amino acids

Amino acid	Three-letter	One-letter
Alanine	Ala	A
Arginine	Arg	R
Asparagine	Asn	N
Asparticacid	Asp	D
Cysteine	Cys	C
Glutamine	Gln	Q
Glutamicacid	Glu	E
Glycine	Gly	G
Histidine	His	H
Isoleucine	Ile	I
Leucine	Leu	L
Lysine	Lys	K
Methionine	Met	M
Phenylalanine	Phe	F
Proline	Pro	P
Serine	Ser	S
Threonine	Thr	T
Tryptophan	Trp	W
Tyrosine	Tyr	Y
Valine	Val	V

Structure of living cells

All living cells are essentially considered to be small balloons full of a complex solution of lipids or oils, proteins, complex carbohydrates or sugars, and nucleic acids. The surface of the balloon, or cellular membrane, is a double layer or bilayer of lipid that contains various proteins floating in the lipid bilayer (Figure 2.1). The cell may be also be surrounded by an extracellular protein or carbohydrate structural layer, such as the cellulose matrix that makes a plant cell rigid. The inner solution of the cell, or cytoplasm, contains proteins, long chains of polymerized chemical residues called amino acids (Figure 2.2, Table 2.1).

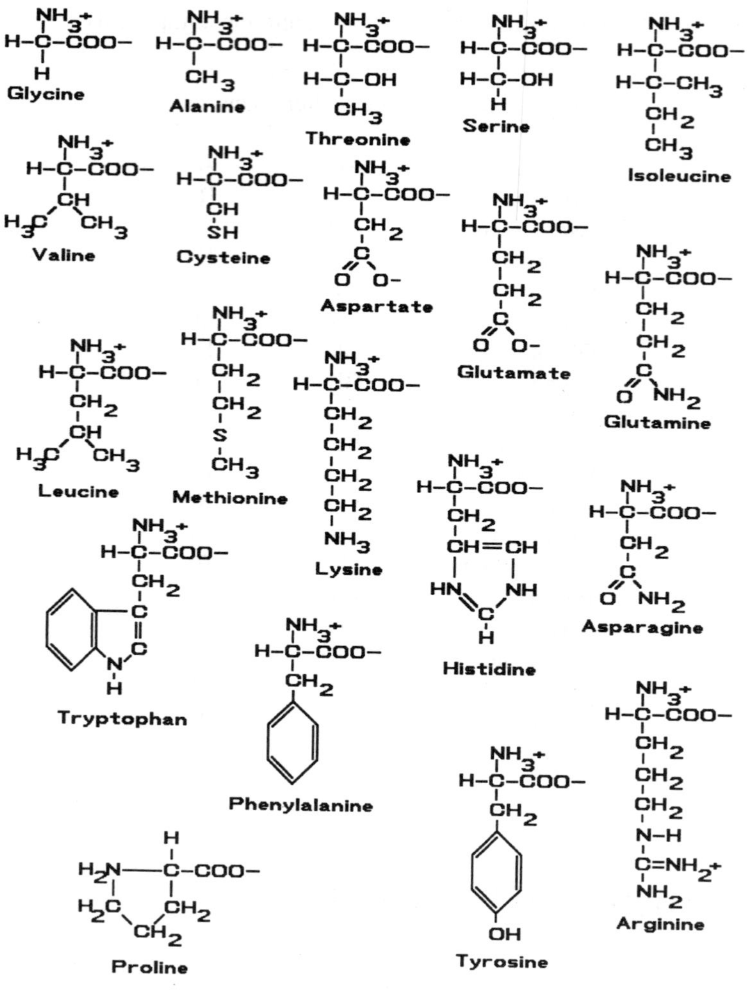

Glycine Alanine Threonine Serine Isoleucine

Valine Cysteine Aspartate Glutamate Glutamine

Leucine Methionine Lysine Histidine Asparagine

Tryptophan Phenylalanine Arginine

Proline Tyrosine

Figure 2.2. Twenty amino acids used for assembling proteins. With the exception of proline, the amino acids used in protein biosynthesis have the common structure NH_2-CHR-COOH, where R indicates a side chain. Under normal cellular conditions, the amino acid has both a positive and a negative charge, indicated as NH_3^+-CHR-COO^-.

Polymerization of amino acids into polypeptide chains

H-N(H)-C(R)(H)-C(=O)-O → H-N(H)-C(R)(H)-C(=O)-N(H)-C(R)(H)-C(=O)-N(H)-C(R)(H)-C(=O)-N(H)-C(R)(H)-C(=O)

Amino acid

Polypeptide chain

α-helix: α keratin

β-pleated sheet: silk (fibroin)

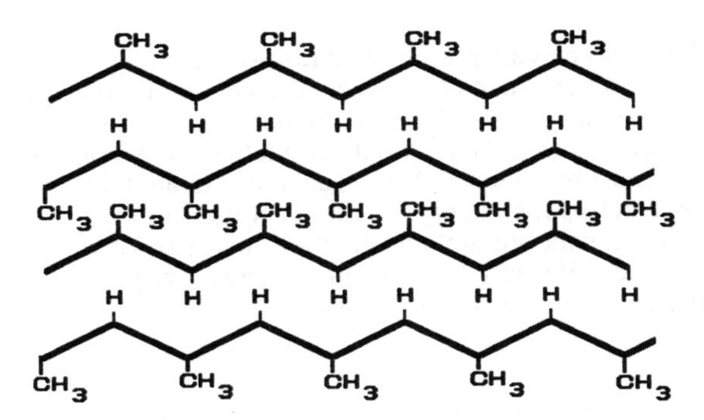

Figure 2.3. Assembly of amino acids into proteins. Proteins are composed of amino acid residues polymerized into linear chains with a backbone of NH-CHR-CO-NH-CHR-CO. The side chains (R) project from the polypeptide backbone and help determine the folding properties of the polypeptide. The polypeptide chain can coil or fold into a distinctive three-dimensional structure that determines the properties of the protein. Two common structures found in proteins are the α-helix and the ß-pleated sheet. In the α-helix, typified by the protein α-keratin, the R groups help force the polypeptide backbone into a coiled helix. In the ß-pleated sheet, found in silk proteins, the alternating alanine (R=CH_3) and glycine (R=H) residues force the polypeptide backbone into an extended zig-zag shape. Most proteins contain a mixture of coils and sheets that determines protein structure and function.

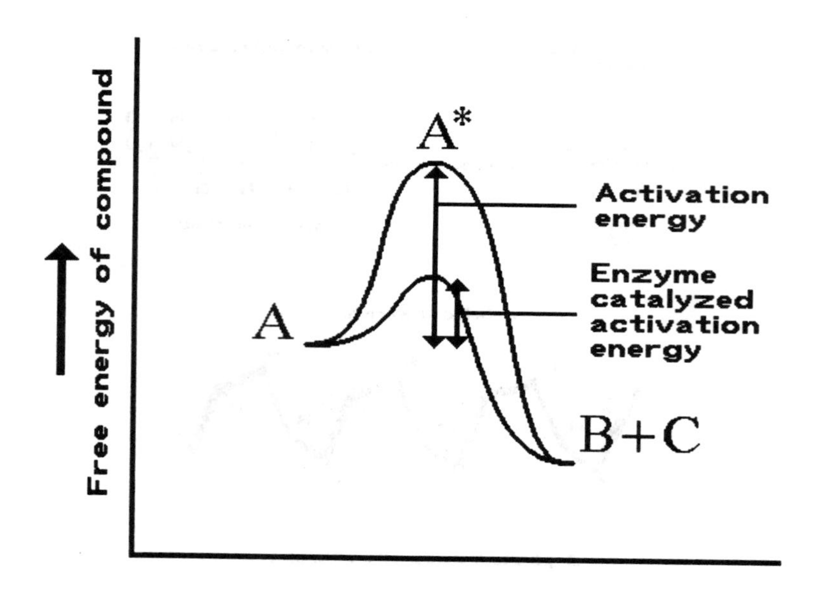

Figure 2.4. Enzymes are proteins that function as catalysts of chemical reactions. In a chemical reaction such as compound A reacting to give the products B and C (A -> B+C), compound A must be activated to a more energetic state (A*) to initiate formation of the products. The energy that must be added to A to create the activated intermediate A* is called the activation energy for the reaction. An enzyme acts as a catalyst to reduce the amount of activation energy required to make the reaction proceed. Because less activation energy is required, the catalyst greatly increases the rate of the reaction.

While many of these proteins have a structural function, like the fibroin protein that makes silk or the keratins that make hair and fingernails (Figure 2.3), other proteins can act as catalysts, or enhancers of chemical reactions (Figure 2.4). These catalytic proteins are called enzymes, and are responsible for cellular metabolism - all of the chemical reactions and biochemical processes that take place within a cell. Since virtually all of the other non-protein components of the cell are actually assembled by enzymes, proteins determine the biochemical properties and the physical characteristics of a cell. Because proteins are the products of the genes present in the chromosomes of a cell, genes are ultimately responsible for maintainance of cellular function.

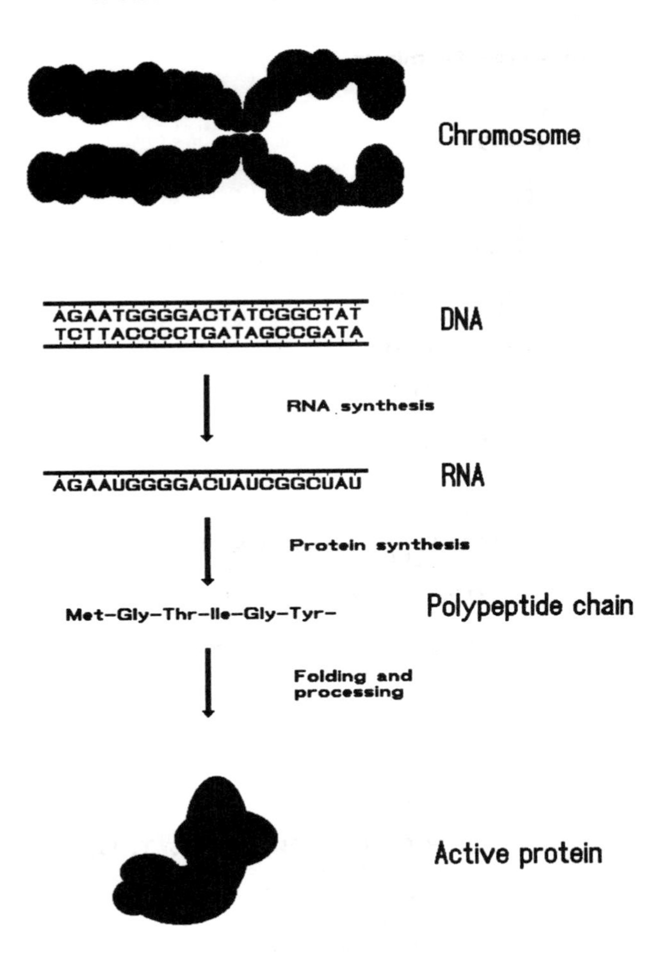

Figure 2.5. Information flow from chromosomes to proteins. Genes that determine the physical characteristics of an organism are present in chromosomes, complexes of many different proteins with the molecule deoxyribonucleic acid (DNA). The genetic information in a chromosome is determined not by the chromosomal proteins, but by the DNA molecule. Genetic information is stored as the order of the chemical residues in the double-stranded DNA molecule. Expression of genetic information requires that the DNA copy of the gene be used as a template or pattern for the synthesis of a ribonucleic acid (RNA) copy of the gene. This RNA molecule can then be used by the cellular protein synthesis machinery for the assembly of a polypeptide gene product. This polypeptide chain must then fold and may undergo processing before becoming an active protein.

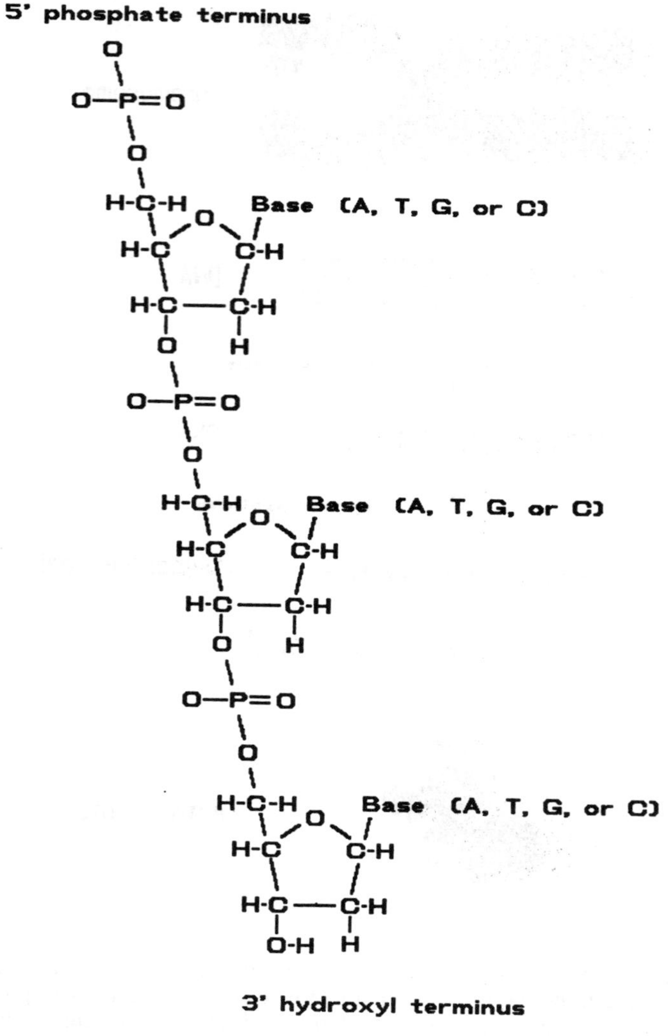

```
5' phosphate terminus
     O
      \
   O—P=O
       \
        O
         \
      H-C-H   O    Base  (A, T, G, or C)
         \ /     \ /
        H-C       C-H
           \     /
         H-C——C-H
           |     |
           O     H
            \
         O—P=O
             \
              O
               \
            H-C-H   O    Base  (A, T, G, or C)
               \ /     \ /
              H-C       C-H
                 \     /
               H-C——C-H
                 |     |
                 O     H
                  \
               O—P=O
                   \
                    O
                     \
                  H-C-H   O    Base  (A, T, G, or C)
                     \ /     \ /
                    H-C       C-H
                       \     /
                     H-C——C-H
                       |     |
                       O-H   H

                    3' hydroxyl terminus
```

Figure 2.6. The chemical structure of the sugar-phosphate backbone of DNA. The sugar deoxyribose has a purine (adenine, A; guanine, G) or pyrimidine (thymine, T; cytosine, C) base attached to the first or 1' carbon. The backbone of the DNA molecule is formed by phosphate bonds that join the 3' hydroxyl group of one sugar to the 5' phosphate of the adjacent sugar. The 5' phosphate and 3' hydroxyl ends of the molecule are indicated.

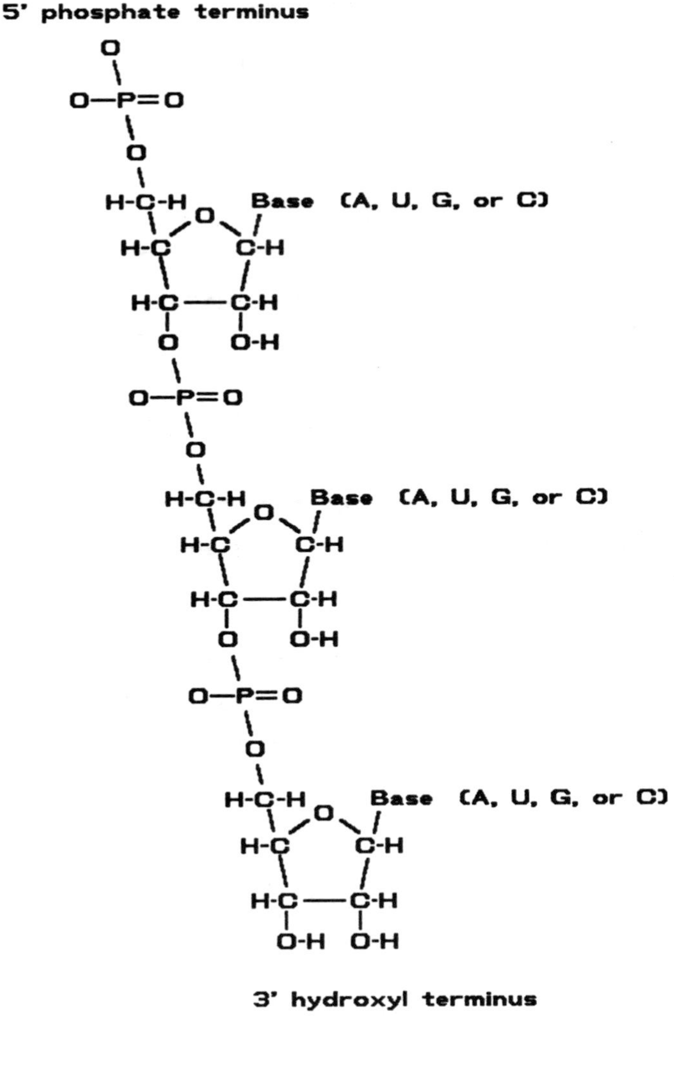

```
5' phosphate terminus
   O
    \
 O—P=O
     \
      O
       \
     H-C-H         Base  (A, U, G, or C)
        \    O    /
        H-C     C-H
           \    /
          H-C——C-H
            |   |
            O   O-H
             \
          O—P=O
              \
               O
                \
             H-C-H         Base  (A, U, G, or C)
                \    O    /
                H-C     C-H
                   \    /
                  H-C——C-H
                    |   |
                    O   O-H
                     \
                  O—P=O
                      \
                       O
                        \
                     H-C-H         Base  (A, U, G, or C)
                        \    O    /
                        H-C     C-H
                           \    /
                          H-C——C-H
                            |   |
                           O-H O-H

                       3' hydroxyl terminus
```

Figure 2.7. The chemical structure of the sugar-phosphate backbone of RNA. The structure of RNA is very similar to that of DNA (Figure 2.6) except that the sugar in the RNA backbone is ribose, which has a hydroxyl (OH) residue attached to the second or the 2' carbon of the sugar. A purine or pyrimidine base is attached to the first or the 1' carbon of the sugar, but RNA contains the pyrimidine uracil (U) rather than the thymine (T) that is found in DNA. The 5' phosphate and 3' hydroxyl ends of the molecule are indicated.

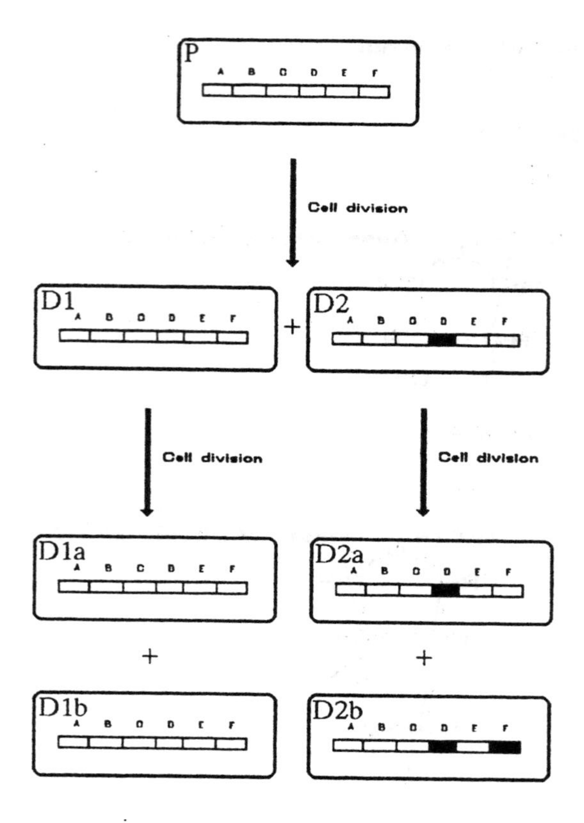

Figure 2.8. Requirement for precise gene duplication during cell division. All of the genes [A through F] in the chromosome(s) of a parental cell [P] must be copied and transmitted to each of the two daughter cells [D1 and D2] during replication and cell division to ensure that the physical traits of the daughter cells are the same as those of the parental cell. The left side of the figure illustrates that error-free gene copying mechanisms yield daughter cells D1a and D1b that are genetically the same as their immediate parent D1, which is in turn genetically identical with its parent P. The right side of the figure illustrates the results of an error-prone gene copying mechanism. After the first division event, one of the daughter cells [D2] has acquired a mutation in gene D and is no longer identical with the parental cell. When this cell D2 divides, one of the daughter cells acquires an additional mutation in gene F. As division proceedes, the genetic characteristics of the daughter cells become increasingly different from the characteristics of the original parental cell.

Deoxyribonucleic acid is used for storage of genetic information

Although proteins are the biochemical catalysts of the living cell, proteins alone do not generally direct the assembly or synthesis of other proteins. The chromosomal theory of heredity postulates that the phenotype, or set of physical characteristics of a cell caused by the particular collection of proteins present in the cell, is controlled by the chromosomes of the cell. Chromosomes are complex structures composed of proteins and nucleic acids. The chromosomal proteins tend to be involved in maintaining chromosome structure and in regulating the expression of genes, the portions of the chromosome that actually carry the information necessary to produce the various cellular proteins. The genes themselves are composed of nucleic acids that store the actual genetic information (Figure 2.5).

Two different types of nucleic acids have been identified as involved in gene structure and function: deoxyribonucleic acid (DNA) (Figure 2.6), and ribonucleic acid (RNA) (Figure 2.7). In general, DNA is used for the long-term storage of genetic information, and RNA is used as a transient or short-lived message to convert the stored information into functioning components of the cell, generally proteins.

During cell division, the chromosomes are duplicated and a set of genes is transmitted to each of the daughter cells (Figure 2.8). It is important that each daughter cell receive an accurate copy of each of the genes present in the parent cell, for if some of the genes were missing or altered in one of the daughter cells, it is likely that the genetically altered daughter cell would have a different phenotype than the parent cell. Although this loss of genetic information might not compromise simple organisms, such as bacteria, complex, multicellular organisms would likely encounter developmental problems if each cell in the organism contained a different set of genes. The chemical structure of the DNA molecule provides a convenient solution to the dilemma of transferring identical copies of genes to daughter cells. The structure of the DNA molecule allows both the

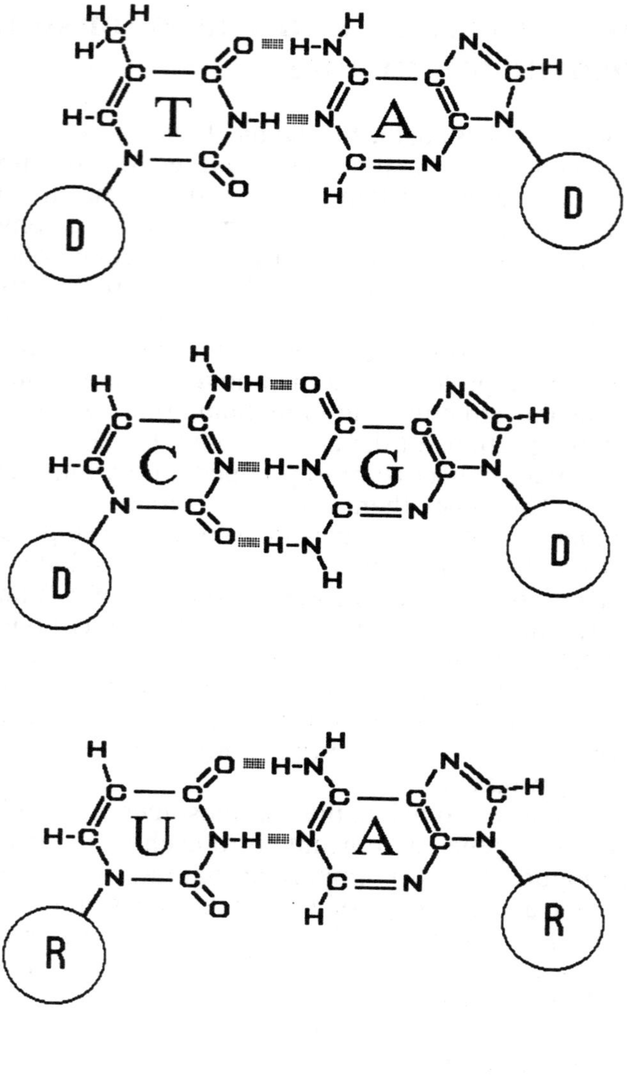

Figure 2.9. The purine and pyrimidine bases attached to the sugar residues in DNA (deoxyribose, D) and RNA (ribose, R) form specific hydrogen bond charge interactions. The purine adenine is able to form two hydrogen bonds, indicated by dashed lines, with the pyrimidine thymine in DNA or the pyrimidine uracil in RNA. The purine guanine forms three hydrogen bonds with the pyrimidine cytosine in either DNA or RNA.

efficient storage of genetic information and the accurate transmission of the information to daughter cells during cell division.

Chemical structure of DNA

DNA is a polysaccharide composed of the five-carbon sugar deoxyribose, phosphate, the purine bases adenine (A) and guanine (G), and the pyrimidine bases thymine (T) and cytosine (C) (Figure 2.9). The cyclic sugar molecules are polymerized into a long backbone by phosphate bonds that connect the fifth or 5' carbon of one sugar molecule to the third or 3' carbon of another. The first or 1' carbon of each sugar is bonded to either a purine or a pyrimidine base. A DNA molecule has a 5' end (generally a 5' phosphate terminus) and a 3' end (generally a 3' hydroxyl terminus), and a string of purine and pyrimidine nucleotide bases attached to the sugar-phosphate backbone. DNA is usually found in the cell as a double-stranded molecule in which the bases of one DNA strand anneal, or pair, with the bases of a second DNA strand to form a ladder-like double-stranded molecule where the rungs of the ladder are composed of the paired bases, and the sides of the ladder are the sugar-phosphate backbone. Note that the two strands of the double-stranded structure have polarity, or 5' to 3' orientation, and that the double-stranded structure is anti-parallel: the 5' end of one strand is paired with the 3' end of the other strand.

To maintain a uniform diameter throughout the length of the double-stranded DNA molecule, it is necessary that the paired bases (the rungs of the ladder) be the same length. This can be accomplished if adenine (A) is only allowed to pair with thymine (T), and guanine (G) is only allowed to pair with cytosine (C). This is possible because of the chemical structures of these bases. A and T are each able to form two hydrogen bonds to stabilize interaction with another base, while G and C are able to form three hydrogen bonds with another base (Figure 2.9).

The anti-parallel structure of DNA, combined with the AT, GC base pairing rule, assures that if you know the sequence of the bases on one strand of a DNA molecule, you also know the sequence of bases on the complementary strand:

```
5'-A-A-T-T-T-C-C-G-G-G-G-T-T-A-A-3'
```

would anneal with the sequence:

```
5'-T-T-A-A-C-C-C-C-G-G-A-A-A-T-T-3'
```

to form the double-stranded molecule:

```
5'-A-A-T-T-T-C-C-G-G-G-G-T-T-A-A-3'
3'-T-T-A-A-A-G-G-C-C-C-C-A-A-T-T-5'
```

Note that is easy to make two exact copies of a double-stranded DNA molecule by separating the two strands and using each single strand as a template, filling in the appropriate complementary bases to produce two identical double-stranded molecules.

```
                              G-G-G-G-T-T-A-A-3'
                            C C-C-C-C-A-A-T-T-5'
                          C   G
                        T   G
           5'-A-A-T-T  A
Parent DNA                                Two identical
           3'-T-T-A-A  T                  daughter DNAs
                        A   C
                          G   C
                            G G-G-G-G-T-T-A-A-3'
                              C-C-C-C-A-A-T-T-5'
```

Messenger RNA is used to convert DNA information into proteins

By 1970, the results of genetic experiments involving many simple organisms including bacteria, yeast, and the fruit fly *Drosophila melanogaster* have led to the hypothesis that genetic information is stored in DNA by means of the linear sequence of the pyrimidine and purine bases in the DNA, but that this information cannot be directly converted into

Single stranded RNA

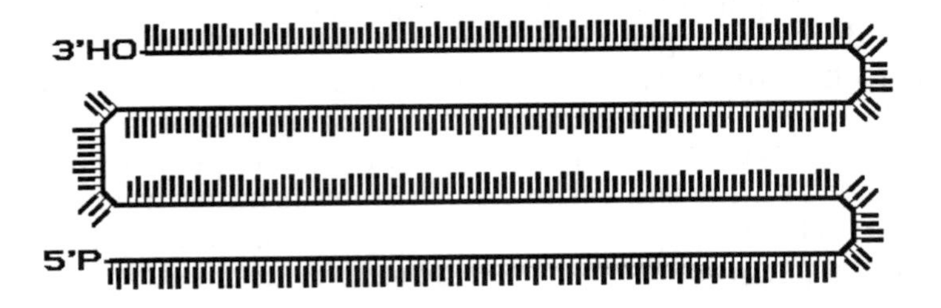

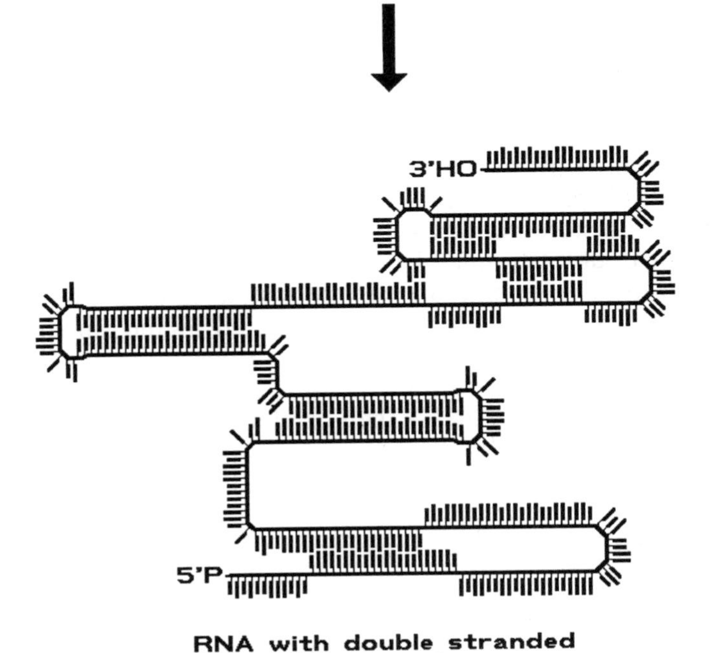

Figure 2.10. RNA secondary structure. Although RNA is frequently indicated as a single stranded nucleic acid sequence, such as: 5'P-AUCGAUAUAGCGCGCUAUAGCUCGAUUUCGCGCGU-3'OH, regions of self-complementarity within the RNA molecule usually allow significant amounts of base pairing to occur, generating regions of double-stranded RNA. This secondary structure may make the RNA molecule resistant to nucleases within the cell or may help fold the RNA into a shape necessary for biological function.

the proteins required in a living cell. Ribonucleic acid (RNA) is part of a complicated protein synthesis system that decodes the genetic information stored in the nucleotide sequences of DNA. RNA is very similar in structure to DNA with several key differences. RNA is composed of the sugar ribose polymerized into a 5' to 3' phosphate polymer with attached purine and pyrimidine nucleotide bases, much like DNA. The purines A and G and the pyrimidine C are present, but the pyrimidine base T is replaced by uracil (U).

Charge interactions and base pairing in RNA are similar to the base pairing found in DNA. As is the case with DNA, G can pair with C, but A pairs with U instead of T (Figure 2.9). In addition, G can form a somewhat weaker interaction with U. While RNA can form double-stranded, base-paired helical structures like DNA, RNA is generally found as a single-stranded molecule with extensive internal base pairing that helps to stabilize the RNA in the cell (Figure 2.10). RNA plays several key roles in the conversion of information stored in DNA to the proteins required by the cell. The double-stranded DNA molecule that constitutes a gene is used as a template for the synthesis of a single-stranded RNA molecule that is assembled by the enzyme RNA polymerase using the AU, GC base-pairing rules. The completed RNA molecule serves as a template, or messenger (mRNA), that is then used for the assembly of the protein encoded by the gene.

The genetic code

The mRNA copy of the DNA gene is used as a blueprint to assemble a protein by using blocks of three mRNA nucleotides at a time as indicators for the correct amino acid residue to be added to a growing polypeptide chain. These nucleotide triplets or codons are read successively in a non-overlapping manner with no spaces allowed between each triplet (Figure 2.11). Since there are four nucleotide residues (A, U, G, C) possible at each position in a triplet, there are $(4)^3$, or 64 possible codons for amino acids (Table 2.2). One codon (AUG) is generally reserved for indicating the start of a protein coding region and three

Table 2.2. The genetic code. The chart indicates the amino acid residue encoded by each of the possible 64 three-nucleotide RNA codons. The amino acid indicated by a specific codon can be determined by finding the 5', middle, and 3' nucleotides of the codon and reading the corresponding amino acid. For example, the codon UUU corresponds to phenylalanine (Phe) and the codon AUG to methionine (Met). The codons UAA, UAG, and UGA do not code for amino acid residues, but are Stop codons that signal the termination point of a polypeptide chain. Note that because of the 5' to 3' polarity of RNA, codons must be read in the 5' to 3' orientation: 5'-AAU-3' codes for asparagine (Asn), but the same codon read backwards as UAA is a stop codon.

5' Base	Middle Base				3' Base
	U	C	A	G	
U	Phe Phe Leu Leu	Ser Ser Ser Ser	Tyr Tyr Stop Stop	Cys Cys Stop Trp	U C A G
C	Leu Leu Leu Leu	Pro Pro Pro Pro	His His Gln Gln	Arg Arg Arg Arg	U C A G
A	Ile Ile Ile Met	Thr Thr Thr Thr	Asn Asn Lys Lys	Ser Ser Arg Arg	U C A G
G	Val Val Val Val	Ala Ala Ala Ala	Asp Asp Glu Glu	Gly Gly Gly Gly	U C A G

(UAA, UGA, UAG) called stop codons or terminators are used to indicate the end of a coding region, leaving 60 remaining codons. As only 20 amino acids are used in the assembly of proteins, the same amino acid may be designated by more than one codon. The code is therefore said to be degenerate. With a few exceptions in certain organisms or in organelles like mitochondria, the genetic

Nucleotide sequence:

ABCDEFGHIJKLMNOPQR

Non-overlapping, non-punctuated triplet code:

ABC DEF GHI JKL MNO PQR

Overlapping, non-punctuated triplet code:

ABC CDE EFG GHI IJK KLM MNO OPQ

Non-overlapping, punctuated triplet code:

ABC EFG IJK MNO

Figure 2.11. Different types of triplet code. Several different types of triplet code might be used to store information in a sequence of nucleotide residues. The non-overlapping, non-punctuated code commonly used in DNA uses three residues per codon and allows no spacer nucleotides between codons. An overlapping, non-punctuated code might use the third residue of one codon as the first residue of the next codon. An overlapping, punctuated code might ignore certain bases. In the example shown, each triplet codon is followed by a single non-coding nucleotide. Although this type of regularly punctuated code has not been detected, the coding regions of eukaryotic genes are frequently punctuated by large regions of non-coding nucleotide residues.

code is considered to be universal and a given triplet codes for the same amino acid residue in all organisms.

As is the case for virtually all aspects of nucleic acid metabolism, the genetic code is always deciphered in a 5' to 3' orientation. However, note that there are three different non-overlapping reading frames for a specific single-stranded nucleotide sequence and a total of six different reading frames for a specific double stranded DNA sequence (Figure 2.12). A DNA fragment can in principle code for a minimum of six different amino acid sequences.

```
Reading frame 3      AsnGlyAspTyrArgLeu→
Reading frame 2     GluTrpGlyLeuSerAla→
Reading frame 1    ArgMetGlyThrIleGlyTyr→

    RNA 1         5'-AGAAUGGGGACUAUCGGCUAU-3'

    DNA           5'-AGAATGGGGACTATCGGCTAT-3'
  fragment        3'-TCTTACCCCTGATAGCCGATA-5'

    RNA 2         3'-UCUUACCCCUGAUAGCCGAUA-5'

Reading frame 1  ←SerHisProSerAspAlaIle
Reading frame 2    ←IleProValIleProXXX
Reading frame 3   ←PheProSerXXXArgSer
```

Figure 2.12. Six potential protein reading frames present in a DNA fragment. A double-stranded DNA fragment can be transcribed into two complementary RNA molecules (RNA 1 and RNA 2). Each of these RNAs can be translated into protein in three reading frames, indicated by the amino acid sequences shown with each RNA. Amino acid residues are indicated using a three-letter code with XXX used to indicate a stop codon. For RNA 1, Reading frame 1 uses nucleotides 123 as the first codon (ACA), Reading frame 2 uses nucleotides 234 as the first codon (CAA), and Reading frame 3 uses nucleotide 345 as the first codon (AAU). Because of the anti-parallel structure of DNA, the reading frames associated with RNA 1 are read from left to right, while the reading frames associated with RNA 2 must be read from right to left.

Transfer RNA as the adaptor molecule for protein assembly

Although the sequence of the triplet codons in an mRNA molecule determines the order of amino acid residues that will be assembled into a protein, the amino acid residues cannot assemble by themselves along the mRNA. Specialized small adaptor RNA molecules called transfer RNA or tRNA (Figure 2.13) are used to align the correct amino acid residues along the mRNA molecule for polymerization by a complex structure called a ribosome.

Each tRNA molecule is only 73 to 93 bases long and

has extensive internal base pairing that folds the molecule into a compact structure with an exposed loop of nucleotides called the "anticodon" at one end of the molecule (Figure 2.14). The anticodon can anneal with an mRNA molecule to form a three-base region of double stranded RNA. Each tRNA contains an amino acid residue attached to the 3' terminus of the RNA molecule. As the tRNA molecules align themselves on the mRNA template, the amino acid residues are correctly positioned to be polymerized to form a polypeptide chain. Each tRNA carries a specific amino acid residue, so these specialized RNA molecules convert the linear sequence of codons present in mRNA to the sequence of amino acid residues that can be assembled into a protein.

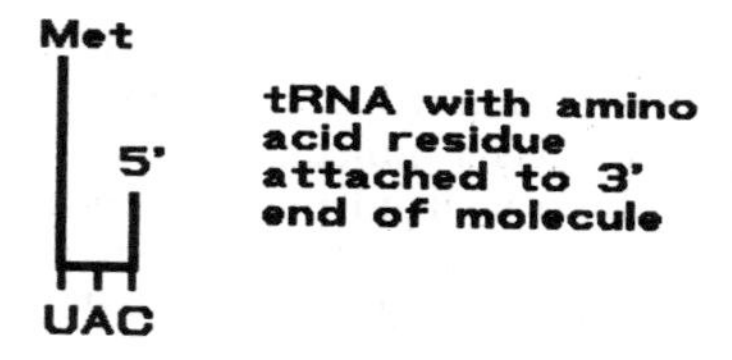

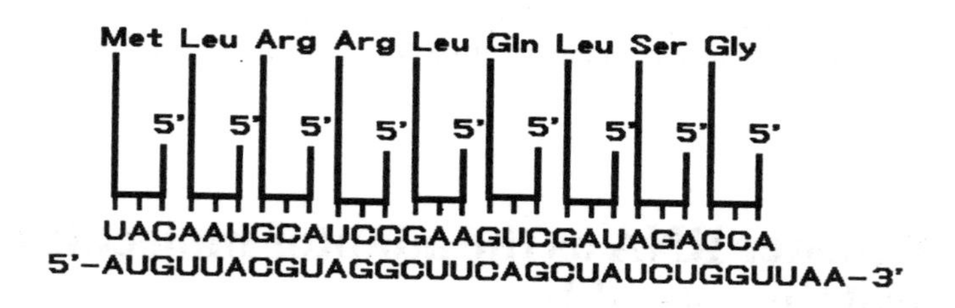

Figure 2.13. Adaptor tRNA molecules with attached amino acid residues decode mRNA and convert the sequence of codons to a sequence of amino acid residues. The anticodon of a charged tRNA molecule can align with the codons in an mRNA molecule to find the complementary codon. For example, the 5'-CAU-3' anticodon of the $tRNA_{met}$ will anneal with the 5'AUG-3' Met codon. When the tRNA molecules are all correctly aligned, it is possible to see the polypeptide that will be produced from the mRNA molecule. The 30 nucleotide RNA in the illustration codes for the 9 amino acid long polypeptide Met-Leu-Arg-Arg-Leu-Gln-Leu-Ser-Gly and the final UAA codon is a terminator that signals the end of the polypeptide chain.

Ribosomal RNA as a structural component of protein assembly

Although the tRNA adaptor molecules are directly responsible for converting the mRNA nucleotide sequence into a sequence of amino acid residues that can be polymerized to form the protein gene product, alignment of the tRNA molecules on the mRNA template does not occur spontaneously. Alignment and assembly requires a complex structure called a ribosome, a collection of many specific proteins assembled together into two subunits, each

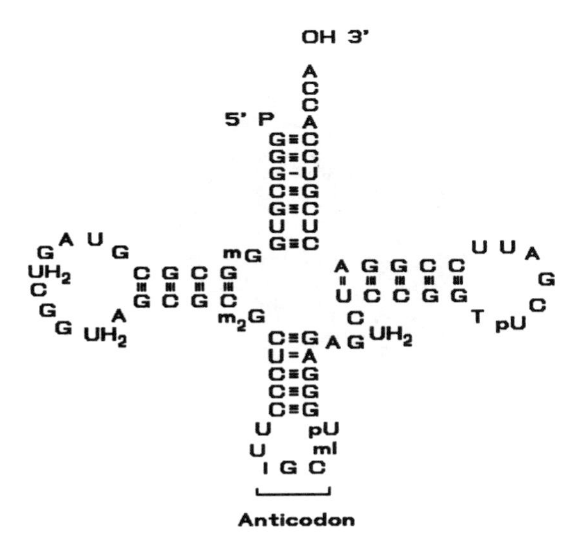

Figure 2.14. Nucleotide sequence and base pairing scheme of the yeast alanine tRNA. Transfer RNAs are small molecules with extensive internal base pairing that is frequently depicted in a cloverleaf structure with the 5' phosphate end base paired with the 3' hydroxyl end of the RNA and three loops stabilized by base paired stems. The anticodon is the portion of the tRNA that aligns with ther complementary codon in an mRNA molecule during protein synthesis. Transfer RNA typically contains several unusual bases including inosine (I), methylinosine (mI), pseudouracil (pU), dihydrouridine (UH_2), ribothymidine (T), methylguanosine (mG), and dimethylguanosine (m_2G).

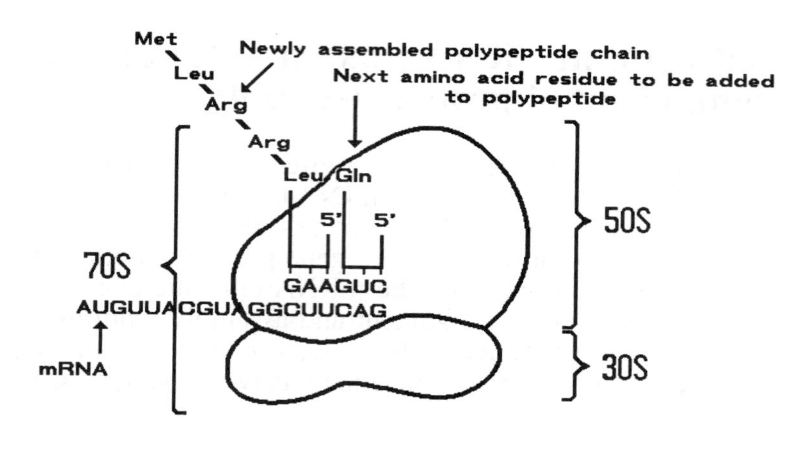

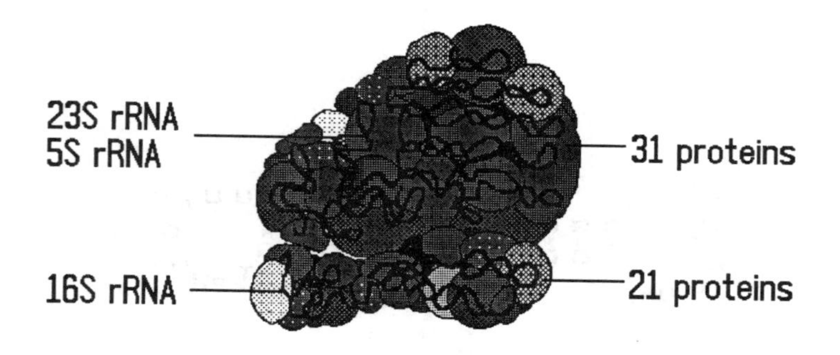

Figure 2.15. Assembly of amino acids into a polypeptide chain takes place on a ribosome. The bacterial 70S ribosome is a complex composed of a large or 50S subunit and a small or 30S subunit. The large subunit contains two RNA molecules, a 23S rRNA of 2904 bases and a 5S RNA of 120 bases, and 31 individual proteins. The small subunit contains a 16S rRNA of 1541 bases and 21 proteins. The RNA molecules can be imagined as forming a framework for the assembly of the proteins into the proper configuration for protein synthesis to occur. The key features of protein synthesis are illustrated by the 70S complex on an mRNA molecule. Transfer RNA molecules with attached amino acid residues align their anticodons with the codons of the mRNA and the ribosome attaches the amino acid residue on the first tRNA to the residue on the second tRNA. The ribosome then translocates, or

containing specific structural RNA molecules (Figure 2.15). Although these structural ribosomal RNAs, termed rRNA, vary slightly in size in different organisms, the overall nucleotide sequence is highly conserved, and is quite similar in all organisms. Eukaryotes generally contain a 25S rRNA, 7.8S, and 5S RNAs in the large ribosomal subunit and an 18S rRNA in the small ribosomal subunit, while prokaryotes contain a 23S rRNA and a 5S RNA in the large subunit and a 16S rRNA in the small subunit.

The reactions catalyzed by the ribosome are extremely complex, but the resultant product of ribosome function is quite simple. Ribosomes use an mRNA molecule as a template to align tRNA molecules in the correct order to connect together or polymerize amino acids into a polypeptide chain. The completed polypeptide can then fold into the correct shape to become a functional protein.

Recombinant DNA and the analysis of genetic information

The overview of cell structure presented above is based on the results of experiments performed by thousands of researchers over decades of work and represents the repeated testing of hundreds of individual hypotheses regarding cell structure and function. The composite view of the structure of the cell illustrates why the order of the

shifts to the next codon on the mRNA, releasing the first tRNA molecule, which no longer contains an amino acid residue. The appropriate charged tRNA molecule then aligns with the next codon and another peptide bond is formed, transferring the growing polypeptide chain to the new tRNA. The ribosome can be imagined as walking down the mRNA molecule in one-codon steps, looking at two adjacent codons at a time and forming a peptide bond between the aligned amino acid residues that are attached to the tRNA molecules. The ribosome shown has the newly synthesized polypeptide Met-Leu-Arg-Arg-Leu attached to $tRNA^{Leu}$, which is aligned with codon 5 of the mRNA. A charged $tRNA^{Gln}$ is aligned with codon 6 ready for the formation of a peptide bond between the Gln amino acid residue and the Leu of the growing polypeptide.

nucleotide bases in DNA and RNA is so important to the function of cellular processes. Genetic information is stored in the chromosome by means of the order of the nucleotide bases in the chromosomal DNA. This information is expressed through the synthesis of complementary mRNA molecules that are assembled using the DNA as a template. The mRNA molecules are then used by the protein-synthesizing ribosomes of the cell to align tRNA molecules in the correct order to assemble the protein product of a gene. All of the components in the cell are either directly the products of genes or are enzymatically assembled by the protein products of genes.

This model of cell structure and genetic function was consistent with the results of hundreds of experiments. The scientific method, however, requires constant testing and re-evaluation of hypotheses. Further testing of the hypothesis that the order of the nucleotide bases in a DNA molecule is the key to the structure and function of genes and gene products necessitated the development of the ability to isolate individual genes and analyze the nucleotide sequences to test whether changes in nucleotide sequence affected gene function.

Recombinant DNA methodology developed as a direct result of research designed to understand gene structure and function by testing the relationship between nucleotide sequence and gene function. As a result, much of this methodology is focused on the isolation, manipulation, and analysis of nucleic acids, the part of the chromosome that stores genetic information.

Summary

1. Genetic information is stored in the chromosomes of organisms. Chromosomes are composed of nucleic acids and proteins.

2. Proteins serve primarily structural and regulatory functions in the chromosome. Genetic information is stored as the sequence of the nucleotide bases present in DNA.

3. The double-stranded, complementary structure of DNA allows accurate production of daughter copies of the genes for transmission to progeny.

4. Information stored in DNA is converted to protein products through the molecule mRNA, which is synthesized using DNA as a template.

5. Determination of the sequence of bases in DNA is important to analysis of gene structure and function.

3

Isolation of DNA

"Everything's got 'em
Everything needs one
Wouldn't be without one
Everything has one." — Nilsson

Because of the need to pass on phenotypic characteristics to offspring, all living organisms must possess a system for storing and transmitting genetic information. With the exception of certain specialized systems that use RNA, such as some viruses, genetic information was by 1970 widely believed to be stored by means of the order of the nucleotide bases present in chromosomal DNA. Testing this hypothesis requires the purification of a DNA fragment that contains a particular gene and the determination of the order or sequence of the nucleotides in the DNA fragment.

This relatively simple goal of isolating and analyzing a DNA fragment must begin with the purification of DNA. Unfortunately, DNA is present in the cell as an association of DNA with many proteins, called a nucleoprotein complex. The DNA must be separated away from these proteins prior to characterization. RNA and polysaccharides can interfere with DNA characterization methods, so the DNA must also be purified away from these macromolecules. In addition, a single cell may contain several different types of DNA molecule. For example, a plant cell will generally contain both mitochondria involved in energy metabolism and chloroplasts involved in carbon fixation. In addition to the DNA present in the chromosomes of the nucleus, both the mitochondria and chloroplasts contain specific, specialized circular DNA molecules that carry genes required for the function of the mitochondria and chloroplasts (Figure 3.1). A simple bacterium, such as *Escherichia coli*, contains only a single, circular chromosomal DNA molecule that carries all the genes necessary for growth and cell division. However, bacteria can also contain extrachromosomal DNA elements, or mini-chromosomes, involved in sexual conjugation or mating, the generation of resistance to antibiotics, or specialized metabolic functions (Figure 3.2). Viruses or bacteriophage that are composed of DNA or RNA may also be present as linear or circular molecules. The analysis of DNA requires not only the purification of DNA away from contaminating proteins and other macromolecules, but often requires the separation of different types of DNA molecules.

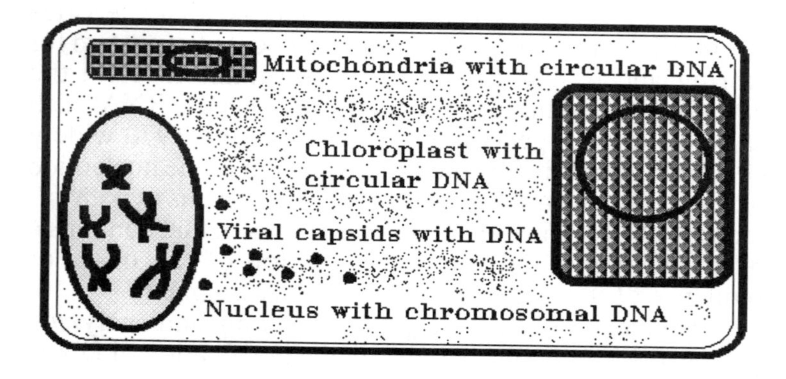

Figure 3.1. Different types of DNA present in a eukaryotic cell. Eukaryotic cells can contain several types of complex structures termed organelles, each of which can contain a different type of DNA. The chromosomal DNA that contains the majority of the cellular genes is present in the nucleus. The mitochondria contain a circular DNA that codes for a few mitochondrial components. Many eukaryotic cells can also contain plastids, which also contain circular DNA molecules. Cells capable of photosynthesis, for example, contain chloroplasts that contain a circular DNA that codes for several components required for chloroplast function. In addition to the DNA present in the organelles, a eukaryotic cell can contain other types of DNA, such as viral DNA or small self-replicating elements.

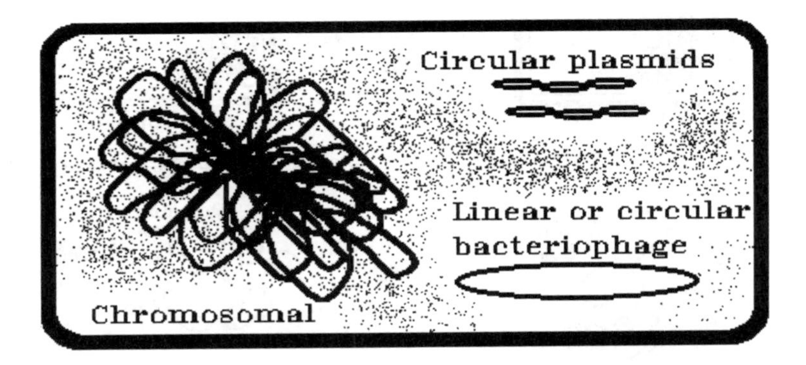

Figure 3.2. Types of DNA present in prokaryotic cells. Although prokaryotic cells do not contain the independent, DNA-containing organelles found in eukaryotic cells, several different types of DNA can be present. The chromosomal DNA is typically present as a large circular DNA molecule organized in a structure called a nucleiod. Many prokaryotes also contain extrachromosomal circular DNA elements like plasmids, F-factors, and resistance transfer factors, sometimes referred to as "mini-chromosomes", which can code for a variety of specialized metabolic functions. Cells can also contain either linear or circular bacteriophage or viral DNA.

Purification of total DNA

The most simple approach to the purification of DNA involves purifying the total DNA present in the cell without regard to separation of different types of DNA molecules (Figure 3.3). This involves rupture or lysis of the cells, removal of proteins, RNA, and other macromolecules, and concentration of the DNA. The lysis solution contains several components to help stabilize the DNA during the purification process. A buffer like Tris-chloride is included to maintain the pH at a constant value, usually around pH 7.5 to 8.0. Since DNA is a long rod-like molecule that can be readily nicked or broken by rough treatment, the solution usually also contains a sugar like sucrose, glycerol, or sorbitol to help stabilize the DNA and minimize the random breaking, called shearing. Since all cells contain proteins called deoxyribonuclease (DNAse) that can degrade DNA, most lysis solutions also contain EDTA (ethylene-

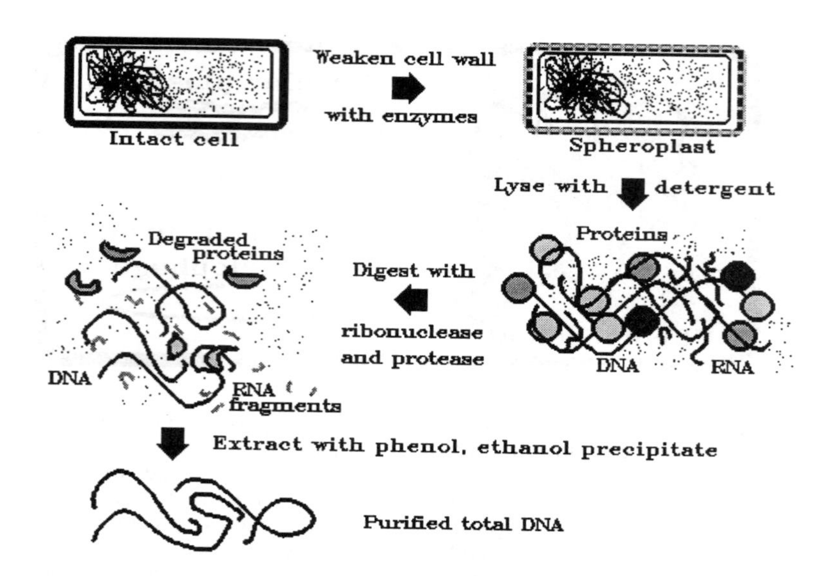

Figure 3.3. Scheme for isolating total DNA from a cell. Cell walls are often weakened by treatment with enzymes and chelating agents like EDTA to remove the metal ions that stabilize cell walls and membranes, then disrupted by the addition of a detergent like SDS (sodium dodecyl sulphate). The crude lysate contains a mixture of DNA, RNA, proteins, and other cellular components. Treatment with ribonuclease and protease partially degrades RNA and proteins and extraction with organic solvents like phenol and chloroform separates the DNA from contaminants. The DNA is finally concentrated by precipitation in 70% ethanol and dissolved in a small volume of buffer prior to further manipulation.

diaminetetracetic acid), a chemical chelating agent that can bind the divalent ions, like magnesium (Mg^{++}), that are required for activity of most DNAses. Lysis solutions may also contain enzymes that degrade cell walls, such as lysozyme. Lysis is usually accomplished by treatment with detergents that disrupt the cellular membranes and release the contents of the cells into the solution.

Following lysis, the solution contains a mixture of DNA, RNA, proteins, and other molecules. A variety of treatments can be used to remove the proteins from the DNA, including the use of enzymes that degrade proteins

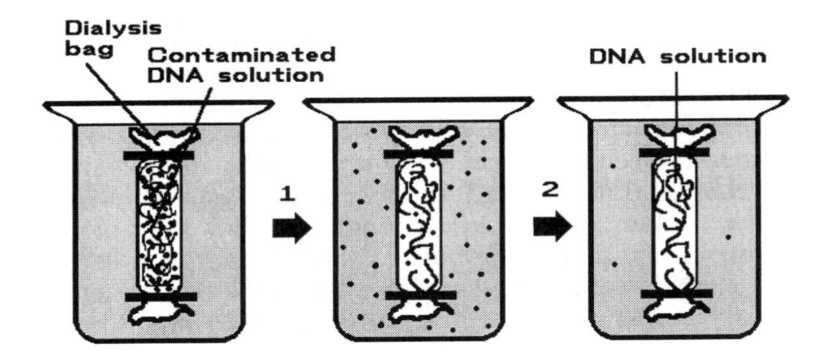

Figure 3.4. Contaminants of low molecular weight, like salts and detergents, can be removed from DNA solutions by dialysis. The DNA solution is placed in a sealed bag made of dialysis tubing, a membrane with small pores that allow small molecules like salts to pass through the membrane while retaining large molecules like DNA. When the sealed dialysis bag containing DNA is placed in a beaker of buffer solution used to maintain a constant pH, the small contaminants pass out of the bag into the surrounding buffer. After a few hours of dialysis (1), the contaminants will be evenly distributed throughout the beaker, greatly reducing the concentration of contaminants inside the dialysis bag. If the dialysis bag is placed into a fresh beaker of buffer (2), the contaminants inside the bag will again diffuse throughout the buffer in the beaker, further reducing the concentration of contaminants in the DNA sample inside the dialysis bag.

(proteases) and extractions with organic solvents (generally phenol and/or chloroform) that denature or unfold proteins and remove them from DNA. RNA that may contaminate DNA preparations is often removed by treatment with enzymes that degrade RNA but not DNA (ribonuclease, RNAse). Following lysis of the cells and extraction of the contaminating macromolecules, the DNA solution is often too dilute for subsequent studies. The DNA is routinely concentrated by precipitation using addition of either two volumes of ethanol or an equal volume of isopropanol. The precipitated DNA can then be dried and resuspended in a small volume of buffer.

Dialysis to remove small molecules

Many small molecules such as salts can contaminate DNA solutions and interfere with subsequent procedures. These small molecules are generally removed by dialysis of the DNA solution (Figure 3.4). Dialysis uses a membrane bag that contains pores large enough to allow salts to pass through the membrane but too small for large molecules like DNA and protein to diffuse out of the bag. When the dialysis bag is placed in a salt-free buffer solution, the contaminating salts diffuse out of the bag into the surrounding solution while the DNA remains in the bag. Each dialysis step dilutes the contaminating salts without affecting the concentration of DNA in the dialysis bag. During dialysis, the contaminating salt will be reduced by the ratio of the volume of DNA solution divided by the volume of the dialysis solution. If 1 ml of DNA solution containing 100 mM contaminating salt is dialysed in 100 ml of buffer, the contaminating salt will be reduced to 1/100 of the orginal concentration (100 mM x 1/100 = 1 mM contaminating salt). Each successive dialysis will reduce the contaminating salt by the same ratio and after three dialysis steps, the contaminating salt concentration will be reduced to .0001 mM (100 mM initial concentration x 1/100 x 1/100 x 1/100). Dialysis reduces the concentration of the contaminating salt to a level that will not affect subsequent work with the DNA.

Cesium chloride density gradient centrifugation

DNA prepared by the simplest approach - lyse, deproteinize, remove RNA, concentrate DNA - can be used for many of the routine procedures of gene analysis. When DNA of exceptional purity is desired, a DNA sample can be subjected to the procedure of cesium chloride density gradient centrifugation (Figure 3.5). A DNA solution is mixed with the heavy salt cesium chloride and placed in tubes in a rotor capable of rotation at speeds of 30,000 to

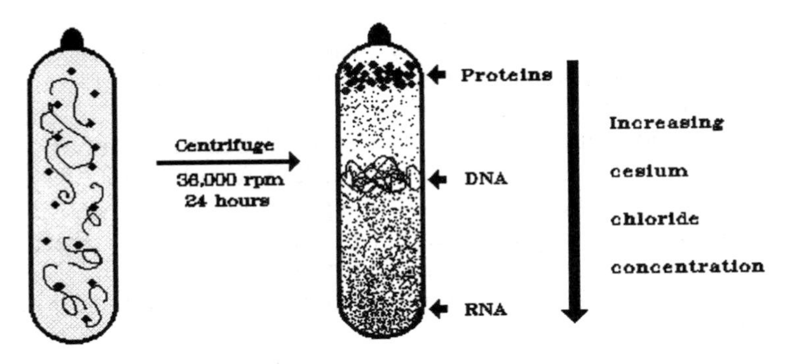

Figure 3.5. Equilibrium density gradient centrifugation for the separation of DNA from RNA and proteins. A solution of DNA, RNA and proteins is mixed with the salt cesium chloride, sealed in a tube, placed in a special rotor, and allowed to spin at 30,000 to 45,000 rpm until the solution reaches equilibrium, typically 24 to 36 hours. At equilibrium, the cesium salt is being forced to the bottom of the tube at the same rate as diffusion is causing the salt to mix throughout the solution. These forces cause a gradient of cesium chloride concentration to develop, with more cesium present at the bottom of the tube than at the top. Molecules placed in such a gradient will tend to form a band in the gradient at a position equal to the bouyant density of the molecules. Since DNA, RNA, and proteins have different bouyant densities, each will band at a different position in the gradient, allowing separation of DNA from RNA and proteins.

70,000 rpm. The rotor is placed in an ultracentrifuge that spins the rotor at high speed in a vacuum. During the centrifugation process, the rotor spins for about 24 hours and the cesium chloride solution establishes a gradient of concentration that is denser at the bottom than at the top. Each of the macromolecules in the tube will tend to float at a concentration of cesium chloride that is equal to the bouyant density of the molecule. This principle can be illustrated with an egg, which will sink in water, but will float in a concentrated salt-water solution. Since DNA, RNA, and proteins have different densities, each of these molecules will band at a different position in the gradient. DNA purified by density gradient centrifugation will be extremely pure and can be used for certain procedures where lower levels of purity could complicate analysis of results.

Cesium chloride density gradient centrifugation can also be used to separate chromosomal DNA into different fractions based on differences in bouyant density of the DNA. Bouyant density of DNA is related to the number of GC base pairs in a DNA fragment. DNA rich in GC base pairs is slightly denser (lower in a gradient) than DNA that is rich in AT base pairs (higher in a gradient), and careful use of density gradients can actually separate GC-rich DNA from AT-rich DNA. A typical eukaryotic genome might contain 50,000 different genes and it is extremely difficult to detect a single GC-rich gene by density gradient centrifugation, since the GC-rich gene would constitute only 1/50,0000 of the total DNA loaded onto a cesium chloride density gradient. Certain DNA fragments, however, may be present several thousand times per genome. The ribosomal RNA genes, for example, are present in several thousand copies, and certain repeated DNA sequences called repetitive elements, such as ATATATATATATATAT, may be present in hundreds of thousands of copies per genome. If these repeated sequences are of sufficiently different nucleotide sequence than the bulk of the chromosomal DNA, they may form a separate, distinct DNA band during cesium chloride density gradient centrifugation (Figure 3.6). These separate bands are referred to as satellite bands because they are distinct from the bulk of the chromosomal DNA.

Purification of circular DNA

Purification of the circular DNA molecules present in extrachromosomal elements, including bacterial plasmids and phage, eukaryotic viruses, and organellar DNA present in mitochondria and plastids, is somewhat more complicated and takes advantage of a difference in the physical properties of a circular and a linear DNA molecule. Chromosomal DNA is generally isolated as large linear molecules, while the DNA in extrachromosomal elements is often present in a closed circular form. If you attempt to twist the helix of DNA, the linear molecule is free to rotate about its central

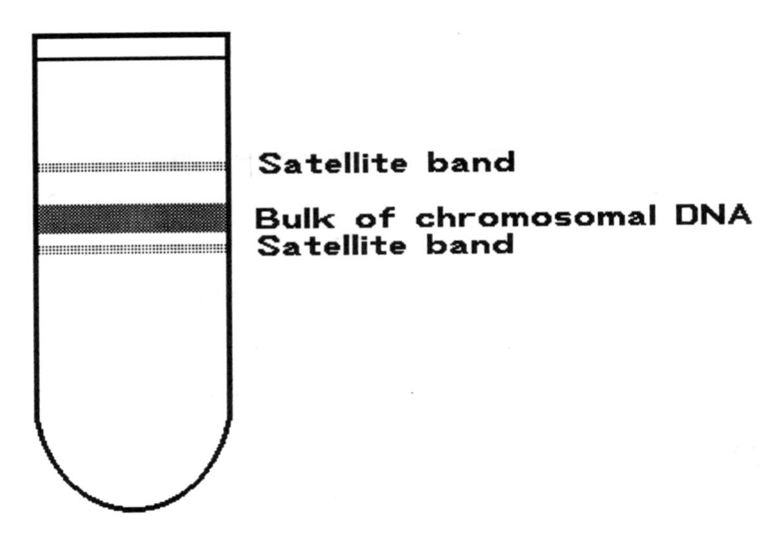

Figure 3.6. Density gradient fractionation of DNA to allow separation of a satellite band from the bulk of the chromosomal DNA. DNA that is rich in GC base pairs is slightly more dense than DNA that is rich in AT base pairs. Although the majority of the chromosomal DNA may form a single broad band on a cesium chloride density gradient, DNA that has a slightly different GC content than the majority of the DNA and that is present in many copies per genome can form separate bands called density satellite bands. The genes that encode the ribosomal RNA molecules and the repetitive DNA sequences associated with chromosomal centromeres, for example, frequently form satellite bands.

axis and can accomodate any twisting strain introduced into the molecule (Figure 3.7). The covalently closed circular, or supercoiled, DNA molecule, however, is topologically constrained and introduced twisting stress cannot be relieved by rotation about the axis of the DNA helix because there are no free ends to rotate. The circular molecule will twist upon itself into a figure eight form, or supercoil, to relieve the stress. Thus, a linear DNA molecule can accomodate more twisting stress than can a covalently closed circular DNA.

Intercalating dyes, such as ethidium bromide and acridine orange, are compounds that bind to DNA by

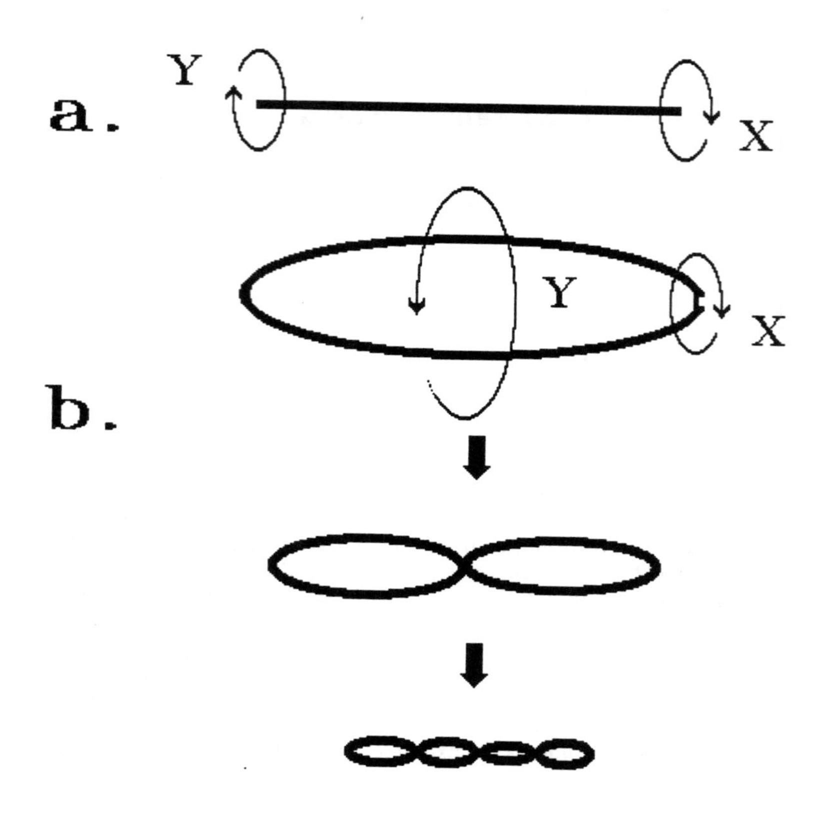

Figure 3.7. Effects of altering the relative twist of a DNA molecule.
(a) If the free end of a linear DNA fragment (position X) is twisted about the axis of the double helix, the stress that is introduced can be relieved by corresponding untwisting of the DNA fragment at the other end (position Y), with no net change in the DNA fragment.
(b) If one strand of a closed circular DNA molecule is nicked (position X) and the nicked strand is twisted around the unnicked strand, the introduced stress cannot be relieved by rotation of the free end of the DNA fragmen t, since a circular molecule has no free end. The introduced twisting stress must be relieved by the formation of superhelical twists in the molecule, and the circle twists around itself (position Y) into a figure eight structure. Increasing amounts of twisting stress will cause the formation of additional superhelical turns until the molecule reaches a maximum level of supercoiling.

inserting themselves between the stacked bases in the center of the DNA helix (Figure 3.8). The binding of each dye molecule is accompanied by an unwinding of the helix to accomodate the strain of the introduced dye. Because closed circular DNA is physically constrained and cannot relieve as much twisting strain as can linear DNA, circular DNA cannot bind as much dye as can a linear DNA of the same size.

Cesium chloride/ethidium bromide density gradients

This propery of intercalating dyes becomes useful in DNA purification when combined with the process of density gradient centrifugation (Figure 3.9). As the ethidium bromide dye binds to DNA, it alters the amount of twist of the DNA molecule and also decreases the buoyant density. Since linear DNA can bind more dye than can circular DNA, the linear DNA will undergo a greater decrease in density than the circular DNA when both molecules are completely

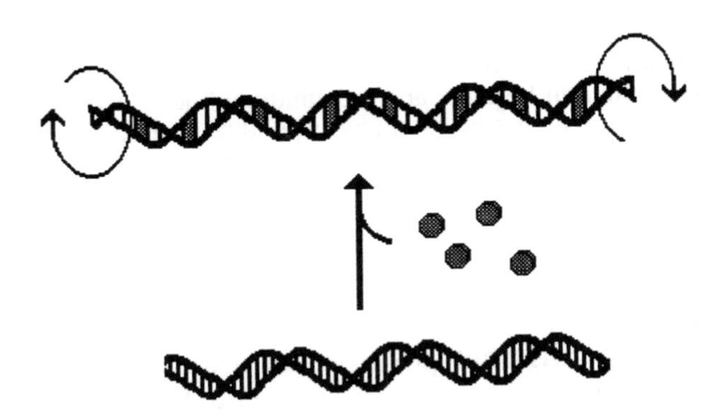

Figure 3.8. Introduction of twisting stress by intercalating dyes. Intercalating dyes like ethidium bromide are flat molecules that insert between the stacked bases of DNA. The inserted dye disrupts the uniform structure of the double helix and the DNA must untwist slightly to accomodate the dye. While a linear molecule is free to rotate about its axis to relieve this twisting stress, a circular molecule is physically constrained, and the binding of ethidium bromide causes the formation of superhelical turns (see Figure 3.7).

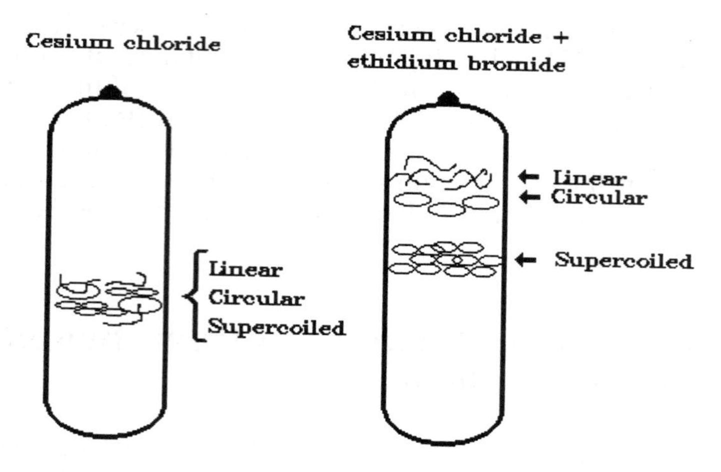

Figure 3.9. Cesium chloride gradients containing ethidium bromide can be used to separate covalently closed circular DNA from linear and nicked circular DNA. Although all three forms of DNA will band at the same position in a normal cesium chloride gradient, the addition of ethidium bromide causes the forms to band in different positions. As ethidium bromide binds to DNA, the bouyant density of the ethidium/DNA complex decreases, and the DNA moves to a less dense position in the gradient (higher up). Because it is free to rotate and relieve the twisting stress as ethidium binds, linear DNA can bind more ethidium bromide than can covalently closed circular DNA, which must form superhelical twists to relieve the twisting stress. The linear DNA can bind more ethidium bromide and the ethidium/linear DNA complex forms a band at a higher position (corresponding to a lighter bouyant density) than does the ethidium/supercoiled DNA complex.

saturated with dye. The dye-saturated circular and linear molecules have different buoyant densities and will form bands at two different positions in a density gradient. Covalently closed circular, or supercoiled, DNA molecules are frequently separated from linear DNA fragments by centrifugation on ethidium bromide-cesium chloride density gradients. Following separation of the circular from the linear DNA, the dye can be removed from the DNA by extraction with an organic solvent like butanol and the cesium chloride removed by dialysis.

Purification of nucleoprotein complexes

Certain types of DNA come "pre-packaged" and can be readily purified as DNA-protein complexes. Bacterial viruses, such as the bacteriophage lambda, contain viral DNA enclosed in a protein coat. The virus can be purified away from the cellular DNA and proteins by cesium chloride density gradient centrifugation, then the viral DNA can be purified away from the protein coat (Figure 3.10). The chromosomal DNA of eukaryotic cells comes packaged in chromosomes in the nucleus. Cells can be lysed in a manner that leaves the nuclei intact and allows purification of the nuclei away from most cellular debris, then chromosomal DNA can be purified by density gradient centrifugation (Figure 3.11). Likewise, chloroplasts and mitochondria can be purified away from nuclei to allow purification of the DNA molecules present in these organelles.

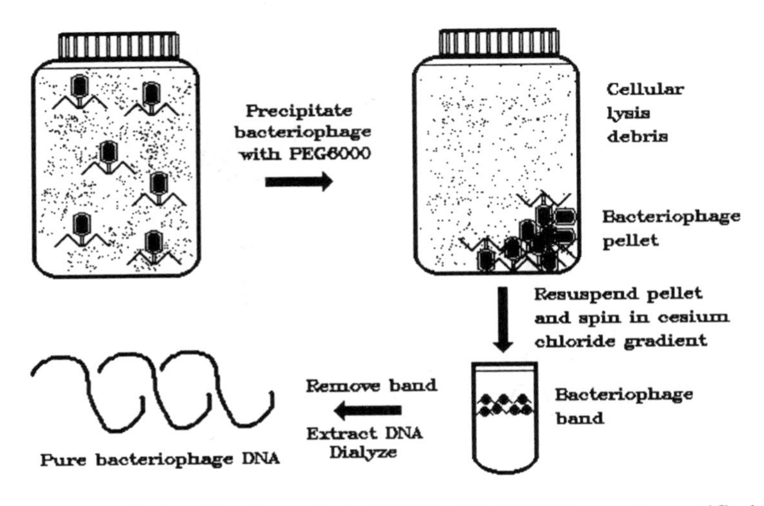

Figure 3.10. Purification of viral DNA. Viral DNA can be purified away from celluar DNA by first isolating the intact virus from an infected culture, then extracting the viral DNA. Purification of the *Escherichia coli* bacteriophage lambda often involves sodium chloride/ polyethylene glycol 6000 precipitation of the bacteriophage from a lysed culture, followed by cesium chloride gradient purification of the bacteriophage. Treatment of the bacteriophage with proteinase and phenol removes the bacteriophage coat proteins and dialysis yields pure bacteriophage DNA.

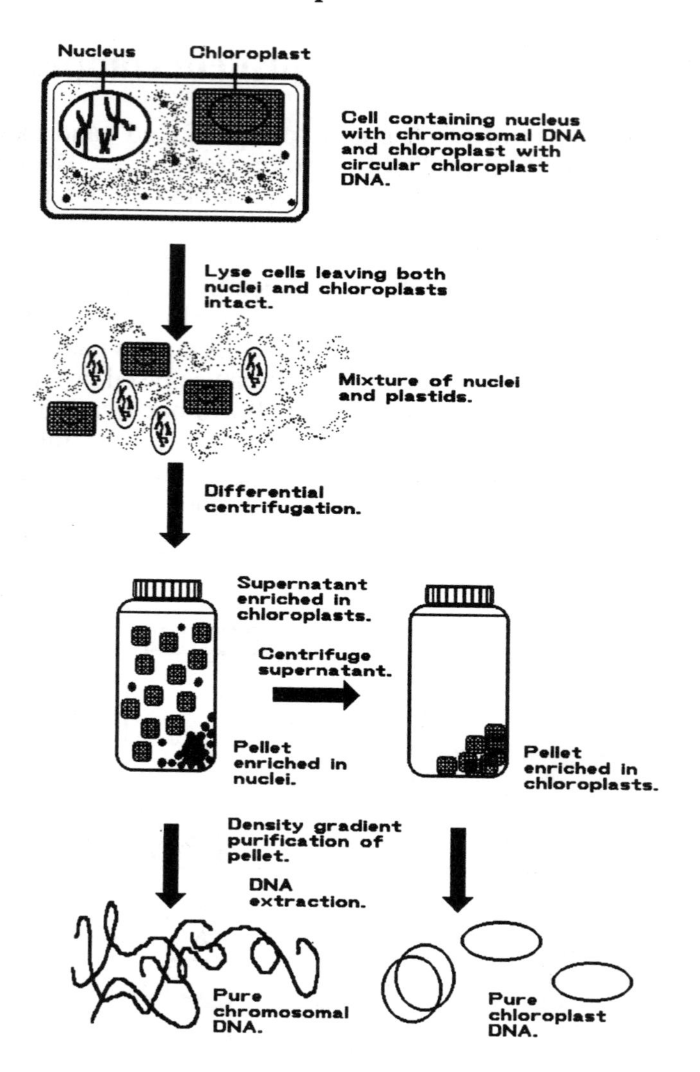

Figure 3.11. The different organelles (nucleus, mitochondrion, chloroplast) present in a eukaryotic cell can be separated prior to extraction of DNA to allow purification of the different types of DNA. Following lysis of the cells under conditions that minimize breakage of the organelles, a low speed of centrifugation can be used to form a pellet enriched in one type of organelle (such as nuclei) and a supernatant solution enriched in a different type of organelle (such as chloroplasts). The nuclei and chloroplasts can be further purified by centrifugation on gradients made of sucrose, Ficoll, or Percoll, and then the DNA extracted.

Quantitation of DNA

Once a DNA sample has been purified of contamination, the amount of DNA present in the sample can be determined using a spectrophotometer - an instrument that measures the amount of light absorbed by a solution. DNA has a characteristic light absorption profile with a maximum at a wavelength of 260 nm (Figure 3.12). A DNA solution of 1 mg/ml will have an absorbance at 260 nm (A_{260}) of 20 optical density units (20 OD). An absorbance of 1 OD at 260 nm therefore represents a DNA concentration of 50 ug/ml [20 OD/1 mg/ml can be divided by 20 to give 1 OD/(1/20) mg/ml]. The ratio of the absorbance at 260 nm divided by the absorbance at 280 nm (260/280 ratio) will generally be 1.8 to 2.0 for pure DNA, although this ratio will vary somewhat with the relative amounts of G/C and A/T in the DNA sample.

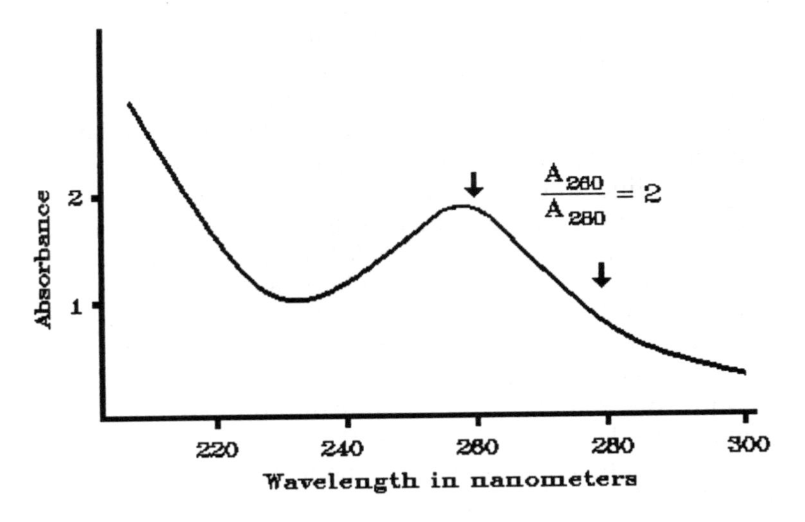

Figure 3.12. Absorption profile of purified DNA. DNA has a distinctive light absorption profile with a maximum around 258 nanometers. Because proteins tend to absorb strongly around 280 nanometers, the ratio A_{260}/A_{280} is frequently used as an estimate of the purity of DNA. Pure DNA has a ratio of 1.9 to 2.0, depending on the nucleotide sequence of the DNA. With increasing protein contamination, this ratio can drop to values close to 1.0.

Miniscreen preparation of DNA

Routine characterization of DNA is in fact not generally performed with extremely pure DNA samples, for extensive purification of many DNA samples becomes time-consuming, expensive, and boring. A variety of protocols called miniscreens have been designed to allow the rapid partial purification of a small amount of DNA from a small number of cells, typically from 1 to 2 milliliters of a liquid bacterial culture. These procedures are commonly used for the simultaneous characterization of many DNA samples. Miniscreens often take advantage of the different physical properties of circular and linear DNA molecules to preferentially discard the linear chromosomal DNA and recover the covalently closed circular forms. Alkaline miniscreens, for example, are frequently used to purify a small amount of circular plasmid DNA from bacterial cells.

Alkaline miniscreens (Figure 3.13) use the ability of a linear double-stranded DNA molecule to denature or melt into two complementary single-stranded molecules when the DNA solution is made very alkaline (pH>10) and generate a denatured form of DNA with different physical properties than covalently closed circular DNA. When a small amount of plasmid-containing bacteria is resuspended in an alkaline solution that contains 1% sodium dodecyl sulfate (SDS), a strong ionic detergent, the cells lyse open and the linear chromosomal DNA fragments denature into single stranded DNA. The complementary strands of the double-stranded circular plasmids, however, are wrapped about one another and are unable to completely separate.

If the solution of DNA is rapidly dropped to a neutral or acidic pH, usually by the addition of sodium or potassium acetate, the denatured strands of DNA will attempt to find their complementary strands and anneal to reform double-stranded DNA. This annealing process is dependent on both time and DNA concentration and each single-stranded DNA molecule can be imagined as having to sort through or try out each of the other single-stranded molecules in the solution in a search to find the correct complementary strand. Because the complementary strands of the closed circular DNA molecules are intertwined around each other,

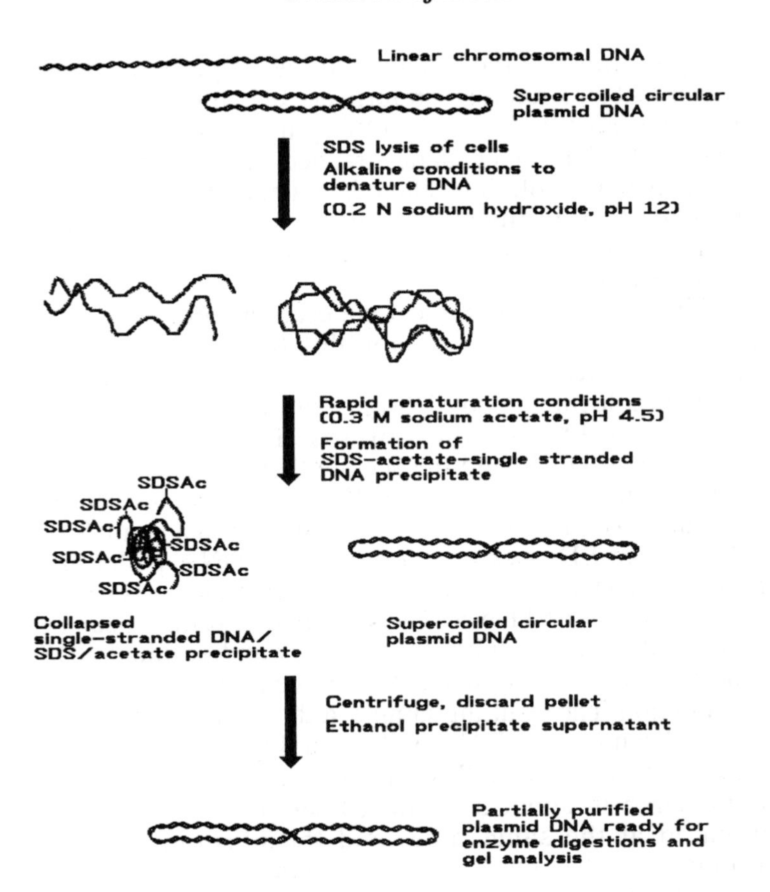

Figure 3.13. Rapid extraction procedures are frequently used for the purification of small amounts of circular plasmid DNA away from linear bacterial chromosomal DNA fragments. Bacteria containing plasmids are lysed in an alkaline solution containing the detergent SDS. Under the alkaline conditions, the linear DNA denatures and separates into single strands. Although the circular plasmid DNA also denatures, the complementary circles are entwined with one another and cannot separate. The denatured DNA sample is then exposed to conditions that force rapid renaturation of the single strands with their complementary strands. Because the single-stranded circular DNAs are entwined with their complements, renaturation is rapid and efficient and double-stranded circles reform. The single-stranded linear fragments are unable to find their complementary strands and collapse in a denatured form, forming a DNA/SDS/acetate precipitate. A brief centrifugation removes this precipitate from the plasmid DNA solution, which can be concentrated by precipitation with ethanol and used for further analysis.

the single strands of the denatured circles rapidly find their complementary strands and reform double-stranded circular DNA. The denatured linear DNA fragments, however, must sort through all the other single-stranded chromosomal DNA fragments in the solution. There is not enough time for the fragments to find their complementary strands and the denatured linear chromosomal DNA fragments collapse into a randomly coiled form.

This denaturation/rapid renaturation step leaves the circular plasmid DNA as double-stranded DNA, but converts the linear chromosomal DNA into collapsed single-stranded molecules. The SDS present in the lysis solution can bind tightly to the collapsed single-stranded chromosomal DNA, and the acetate that was added to the solution interacts with the SDS to form an insoluble precipitate. A 5 minute spin of the solution in a centrifuge will form an SDS-acetate-chromosomal DNA pellet and the solution on top of the pellet, the supernatant, will contain the circular plasmid DNA. The supernatant solution can be poured into a clean tube, the proteins and residual SDS removed, and the circular DNA concentrated by alcohol precipitation. The resulting sample, although not completely free of chromosomal DNA, will be highly enriched for the circular plasmid DNA. The miniscreen DNA sample can be prepared in about 2 hours, 18 samples can easily be prepared at the same time, and the plasmid DNA is sufficiently free of chromosomal contamination for further characterization.

Summary

1. **DNA present in cells is associated with many proteins.**
2. **DNA must be purified away from other cellular macromolecules prior to analysis of the DNA.**
3. **DNA can exist in different forms or conformations with different purification properties.**
4. **DNA can be quantified by its absorbance of light at wavelength 260 nm.**
5. **Small samples of DNA can be rapidly prepared using miniscreen procedures.**

4

Specific Cleavage of DNA

"I've never really been lost, but there have been times when I didn't know where I was for a couple of days." — attributed to Daniel Boone

Identification of a point of reference on a DNA molecule as it is isolated from a cell is a technical innovation that was crucial to the development of recombinant DNA methodology. Chromosomal DNA purified from cells is frequently obtained as a randomly nicked or sheared population of fragments of 30,000 to 200,000 base pairs in size. In principle, each fragment has different ends than every other fragment in the population and no two fragments are identical. If you can imagine yourself standing on one of these fragments, since all of the fragments look alike, it can be difficult to determine precisely which DNA fragment lies under your feet. Some sort of molecular reference point is necessary to determine the physical location of genes on a DNA molecule.

As an additional complication, the fragments of DNA are not even the same size and, because the fragments are generated through random breakage, a desired gene may be present on fragments of many different sizes. A typical gene of interest might be 3000 base pairs long, while the purifed DNA preparation containing the gene might be an average size of 50,000 base pairs, with each fragment different from every other fragment (Figure 4.1). Some fragments will contain the desired gene, others will contain a part of the gene, and many fragments will not contain any of the desired gene sequence. It is important to be able to define or identify reference points on the DNA molecules to generate a map so that the molecules can be aligned and compared with one another.

Genetic maps

Conventional genetic approaches have been used to establish genetic maps for many organisms. Genetic maps are constructed using the assumption that during

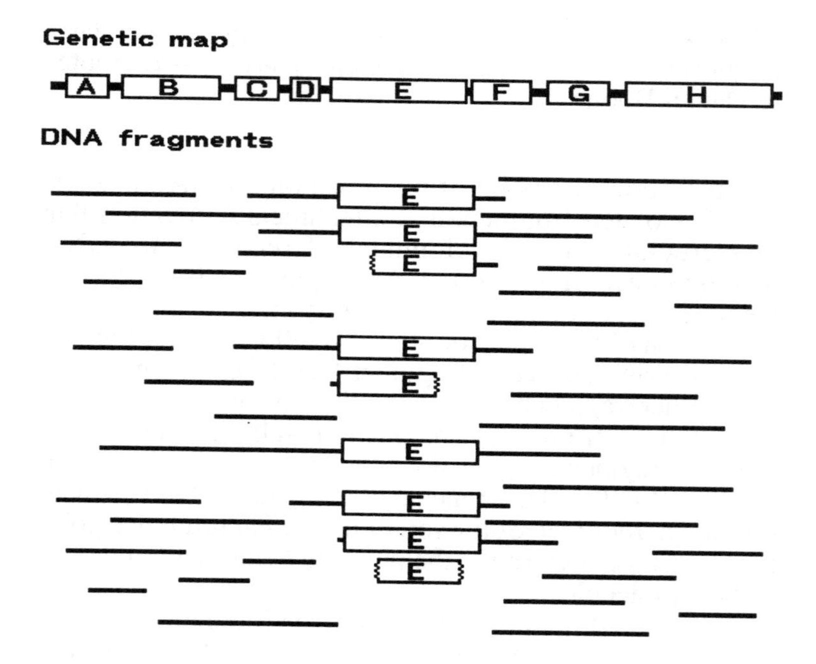

Figure 4.1. Random sizes of purified chromosomal DNA fragments. The purification of chromosomal DNA causes random breaks that fragment the DNA into pieces with an average length of 50,000 to 200,000 base pairs. Because these breaks occur randomly, an individual DNA fragment may contain all of, a portion of, or none of a desired gene. The chromosome illustrated contains eight genes (**A**, **B**, **C**, **D**, **E**, **F**, **G**, and **H**) and portions of gene **E** are present on both large and small DNA fragments. DNA fragment size cannot be used as a method for identifying or isolating gene **E**.

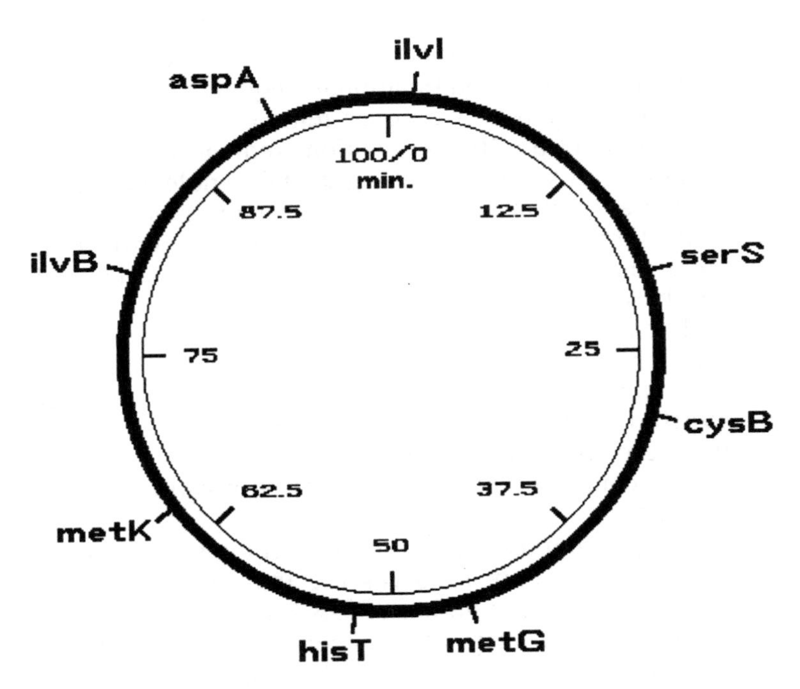

Figure 4.2. Genetic map of *E. coli*. The relative positions of eight genes involved in amino acid metabolism are shown on the circular bacterial genetic map. The units of measurement are minutes, a measure of the time required for transfer of each gene during conjugation. The ilv*I* gene is transferred very early, transfer of met*G* requires about 40 minutes, and asp*A* requires about 90 minutes for transfer. Genetic maps can also be displayed in units that depict recombinational distance between genes.

chromosomal recombination, two genetic markers that are located close to each other on the chromosome are more likely to remain together than two genetic markers that are located far apart on the chromosome. By using recombination frequency as a measure of the order and relative position of genes, genetic maps can be constructed to depict the genes that are present on a DNA molecule (Figure 4.2).

Several problems complicate the direct correlation of a genetic map with the actual sequence of a DNA molecule. The map units used to depict the positions of the genes are generally based on recombinational frequency, a measure of the probability that two genes remain together during sexual reproduction. Recombination frequency, however, is not necessarily uniform along a DNA molecule. Two genes that are 1 map unit apart on a genetic map may be located 20,000 base pairs from each other, while two different genes that are also 1 map unit apart may be located 45,000 base pairs from one another. It can therefore be difficult to make a direct correlation between a genetic map and the actual sequence of a DNA molecule. In addition, a genetic map can only be made as complete as the number of available chromosomal mutations that alter the phenotype, or appearance, of the organism. Genes for which mutations cannot be obtained cannot be positioned on a genetic map.

Perhaps the most perturbing problem in generation of a physical DNA map involves the inability to align a randomly sheared population of DNA fragments with an established genetic map, as has been done in Figure 4.2. There is no easy way to physically look at each individual DNA fragment and determine which portion of the chromosomal genes are present on that particular DNA fragment. The ability to construct a physical map of a DNA molecule and correlate or compare this physical map with a genetic map is of fundamental importance to analysis of gene structure and function.

Site-specific cleavage of DNA

The discovery of proteins that digest or cleave DNA in a site-specific manner provided the basis for the generation of a physical restriction map based not on the genetic characteristics of DNA, but on the actual sequence of the nucleotide bases in DNA. These enzymes, called restriction endonucleases, are proteins that recognize specific sequences of nucleotide base pairs in the DNA molecule. Cleavage sites are typically sequences four to eight base pairs in length and often possess dyad, or two-fold rotational, symmetry. *Eco*RI, the enzyme purified from *E. coli* strain RY13, recognizes and cleaves at the dyad sequence

```
          plane of symmetry
                  |
       5'-NNNGAA | TTCNNN-3'
       3'-NNNCTT | AAGNNN-5'
                  |
```

where the six-base sequence GAATTC occurs in a stretch of other nucleotides (N) and the plane of rotational dyad symmetry is indicated. The enzymes are named after the organism from which they are purified. The naming system uses the form of *Genus species* (strain) (# of enzymes in the strain). The underlined letters form the name of the enzyme, the strain is usually indicated as a letter, and a Roman numeral indicates which enzyme when more than one is present in that organism. *Hin*dIII, for example, is one of three enzymes found in *Hemophilus influenzae* strain d. When two enzymes purified from different sources are found to cleave at the same site, the enzymes are called isoschizomers of one another.

The frequency of occurrence of a restriction enzyme cleavage site in DNA can be mathematically estimated by the formula

$$\text{Frequency} = (1/4)^n \text{ base pairs}$$

where n equals the number of bases in the recognition site and there are four possible bases (A,C,G,T) that might occur at each of the positions in the recognition sequence. The 2-base sequence AT would occur at a frequency of $1/4^2$ base pairs, or once in 16 base pairs, the 4-base sequence GATC would occur once in 4^4, or 256 base pairs, and the 6-base sequence GGATCC would occur once in 4^6, or 4096 base pairs. The frequency of occurrence of the cleavage site decreases as the size of the recognition site increases.

Some restriction enzymes recognize sites that are called degenerate because the enzymes will recognize two different but closely related nucleotide sequences. The enzyme *Hin*fI, for example, recognizes the 5-base sequence GANTC where N can be either A, C, G, or T. Although this sequence is 5 bases long, the frequency of occurrence will be 1/4 x 1/4 x 4/4 x 1/4 x 1/4, or $1/4^4$ instead of $1/4^5$, because any of the four bases can occur in the third position of the sequence. The degree of degeneracy in a sequence affects the frequency of random occurrence of the sequence in DNA. Allowing either of 2 bases at a particular position in a sequence doubles the frequency of the sequence, while allowing any of the 4 bases in a position increases the frequency by a factor of four (Table 4.1).

Table 4.1. Effect of degeneracy on the frequency of occurrence of restriction enzyme recognition sites in DNA. The size, degeneracy, and frequency of occurrence are shown for a 4-base and four 5-base sequences with different degrees of degeneracy. Degeneracy in the sequences is indicated by N=any of the four bases, Pu=either of the purines (A or G), and Py=either of the pyrimidines (C or T).

Sequence	Size in bases	Degeneracy	Frequency in base pairs
GATC	4	0	1/256
GAPuTC	5	2-fold	1/512
GAPyTC	5	2-fold	1/512
GANTC	5	4-fold	1/256
GAATC	5	0	1/1024

Estimation of the frequency of occurrence of a sequence is based on the assumption that the order of bases in DNA is random and this is clearly not true. Because DNA has the important biological function of coding for or storing the information for all of the RNA and proteins required by the cell, there are many constraints or restrictions on the order of the bases in DNA. The actual frequency of restriction site occurrence does not always match the predicted frequency of occurrence. It is possible, for example, to find a 20,000 base DNA fragment that contains no cleavage sites for *Eco*RI (GAATTC) although 4-5 sites are expected from the calculated frequency of occurrence (1/4096 base pairs). The actual frequency of occurrence can be either higher or lower than the calculated frequency.

Types of end generated by restriction enzymes

Restriction enzymes can cleave DNA to generate two types of termini: blunt and cohesive. If an enzyme that recognizes a six-base sequence cleaves between the third and fourth bases, the resulting terminus is flush or blunt-end:

```
5'-NNNNNNNNNGAA   TTCNNNNNNNNNNN-3'
3'-NNNNNNNNNCTT   AAGNNNNNNNNNNN-5'
```

If the cleavage is between the first and second bases of the sequence, the result is a four-base, cohesive, complementary terminus with a 5' extension:

```
5'-NNNNNNNNNG       AATTCNNNNNNNNN-3'
3'-NNNNNNNNNCTTAA       GNNNNNNNNN-5'
```

Cleavage between the fifth and sixth bases generates a similar terminus with a 3' cohesive extension:

```
5'-NNNNNNNNNGAATT       CNNNNNNNNN-3'
3'-NNNNNNNNNC       TTAAGNNNNNNNNN-5'
```

Some restriction enzymes, such as the class IIs enzymes, do not recognize and cleave within symmetrical sequences, but cleave the DNA some distance to one side of the recognition sequence. *Hga*I, for example, recognizes the five base sequence:

```
5'-NNNNNNNNNGACGCNNNNNNNNNNNNNNNNNNNN-3'
3'-NNNNNNNNNCTGCGNNNNNNNNNNNNNNNNNNNN-5'
```

Cleavage occurs five bases to the right on the top strand

and ten bases to the right on the bottom strand to generate a five-base cohesive terminus that is fairly specific for the DNA fragment cleaved:

```
5'-NNNNNNNNNGACGCNNNNN          NNNNNNNNNNNNNNN-3'
3'-NNNNNNNNNCTGCGNNNNNNNNNN          NNNNNNNNNN-5'
```

Inhibition of restriction by modification enzymes

Restriction enzymes not only cleave DNA *in vitro*, or in the test tube, but also *in vivo*, in the viable cell. At first glance, this would appear to be a suicidal situation for the cell, for a cell that contains a restriction endonuclease should digest its own chromosomal DNA. However, the cells that make restriction endonucleases also make enzymes called modification methylases that can prevent the cleavage of DNA by the corresponding restriction enzyme. These methylases modify bases within the cleavage site of the restriction enzyme, generally by attaching methyl groups to certain positions of A or C residues. This methylation destroys recognition by the restriction enzyme, thus protecting the cell's own DNA.

During DNA replication, the newly sythesized DNA is unmethylated, but the parental strand of the DNA molecule is likely to be methylated, so that the restriction cleavage sites in newly synthesized chromosomal DNA tend to be modified on one of the two strands of the DNA. This hemi- or half-methylated DNA is not generally a good substrate for restriction enzymes, so the cellular DNA is resistant to digestion by its own restriction enzymes. Any unmodified DNA that enters the cell, however, is subject to cleavage and degradation. Restriction enzymes are believed to play a role in restricting the entry of foreign DNA's, such as infecting viral DNA.

Genetic map

EcoRI enzyme cleavage sites in DNA fragments

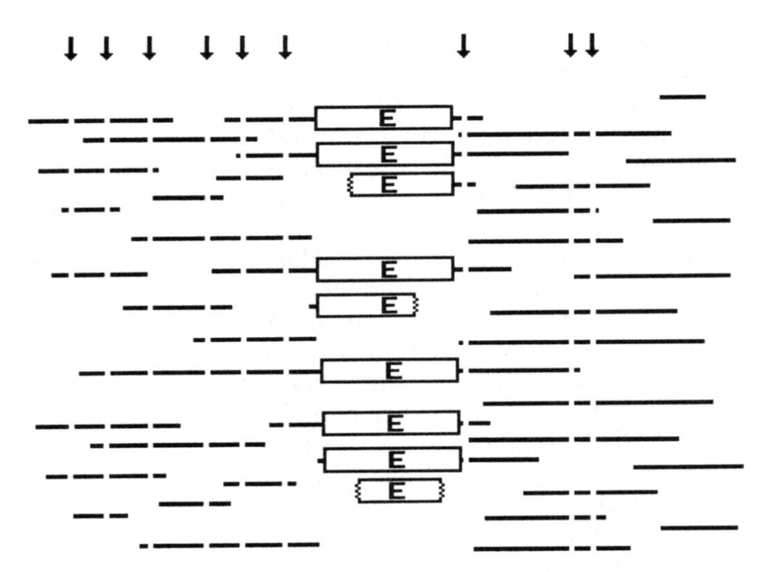

Figure 4.3. Digestion of a randomly nicked chromosomal DNA sample with a restriction endonuclease generates DNA fragments of specific sizes. Restriction enzymes cleave at specific nucleotide sequences within the DNA molecules. When the random population of chromosomal DNA fragments shown in Figure 4.1 is treated with restriction endonuclease *Eco*RI, the enzyme will cleave each time the nucleotide sequence GAATTC occurs in a DNA molecule, indicated by the arrows above the DNA fragments. Digestion generates a new set of smaller DNA fragments of two different types: specific fragments with an *Eco*RI site at each end and non-specific fragments with an *Eco*RI site at one end and a random break at the other end. When there are more specific than non-specific DNA fragments, the specific fragments will form discrete size classes. In the illustration above, gene **E** is always present on the largest specific *Eco*RI fragment. Portions of gene **E** will be present on smaller non-specific DNA fragments with an *Eco*RI site at one end and a random break at the other.

Restriction enzymes generate specific DNA fragments

Digestion of a purified chromosomal DNA sample with restriction enzymes overcomes the problem of the heterogeneous mixture of DNA fragments normally present in a chromosomal DNA sample. Restriction enzymes cleave DNA at specific positions that are determined by the order of the nucleotide bases in the DNA molecule. When the randomly sized chromosomal DNA fragments are completely digested with a restriction enzyme, the DNA will be cleaved at all of the recognition sites for that enzyme. This cleavage process generates a subset of smaller DNA fragments derived from the original large, randomly sheared fragments (Figure 4.3). Two distinct types of smaller fragments occur: specific fragments where both ends are the result of restriction enzyme cleavage and non-specific fragments with a cleavage site at one end and a randomly sheared terminus at the other end. Each large, randomly sheared DNA fragment gives rise to 3 to 5 specific and 2 non-specific fragments. Since the starting chromosomal DNA population contains thousands of different DNA fragments, thousands of copies of each of the specific fragment types accumulate in the digested DNA sample to form specific size classes of fragments. The non-specific fragments also accumulate in the digested DNA sample, but since these fragments are of many different sizes, specific size classes do not accumulate. This is perhaps the single most valuable feature of restriction enzymes. Digestion of the randomly sized chromosomal DNA with a restriction enzyme generates specific sets of DNA fragments that are defined at their ends by the restriction cleavage sites.

Separation of DNA fragments by size

Following digestion of a DNA molecule with restriction enzymes, the size classes of DNA fragments must be separated from one another so that the DNA fragments can be visualized and the sizes of the fragments determined. Separation of DNA fragments according to size is generally accomplished by gel electrophoresis. The phosphate molecules that are present throughout the backbone of DNA have a strong negative charge and when DNA is placed in an electric current, the DNA fragments will migrate towards the positive terminus. In the absence of anything to impede the migration of the fragments, fragments will all migrate at about the same rate, because the charge-to-size ratio for DNA is constant (a molecule of twice the size will have twice the charge, but also twice the mass to move).

Electrophoresis is performed in a matrix, such as agarose or polyacrylamide, that resists the migration of the DNA fragments by forcing the fragments to work their way through small pores in the matrix (Figure 4.4). The matrix can be imagined as a fine mesh or grid through which the DNA molecules must filter in order to move with the electric current. Because it is easier for smaller fragments to weave through the grid than it is for larger fragments, the DNA fragments will separate according to size, with small fragments migrating much faster than large fragments. When a DNA sample is loaded at one end of a gel in a small slot called a sample well, fragments of different sizes will resolve into bands as electrophoresis proceeds (Figure 4.5).

A gel must contain sufficient ions to allow electric current to flow, and must be buffered to prevent changes in pH that occur when ions move during electrophoresis. Gel buffer systems often contain the compounds Tris, EDTA, and boric or acetic acid at a pH of about 8. DNA bands are most often visualized by staining of the gel (either during or after electrophoresis) with ethidium bromide, which binds to the DNA fragments. Exposure to ultraviolet light then causes the DNA fragments to fluoresce, allowing routine detection of as little as 2 nanograms (2×10^{-9} gram) of DNA in a band. Because ethidium bromide and UV light can both act as mutagens to introduce changes in DNA, some element of caution is necessary in handling the ethidium solutions. The less hazardous, but also less sensitive, dye methylene blue can also be used to stain DNA bands in gels.

Agarose and polyacrylamide gel matrices are useful

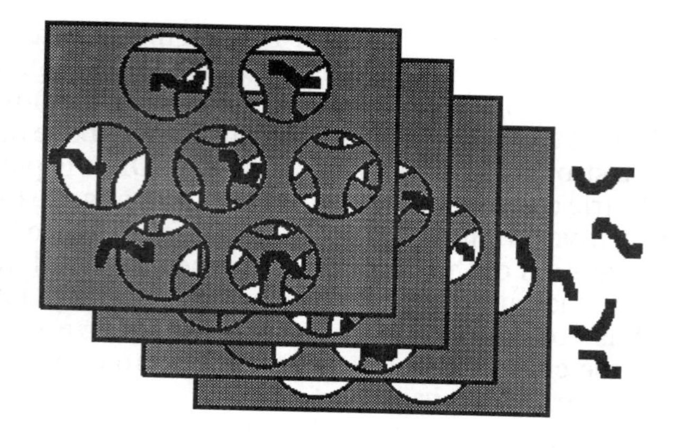

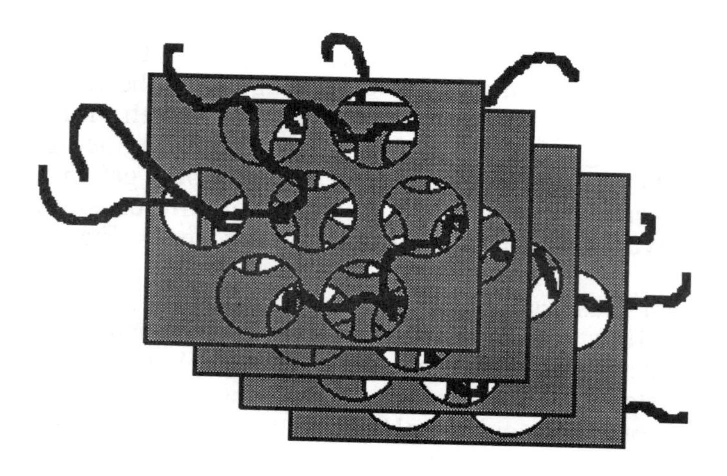

Figure 4.4. Principles of gel electrophoresis. Because the strongly negative charge on the sugar-phosphate backbone of a DNA molecule gives DNA fragments a fairly constant charge-to-mass ratio, fragments of different size that are placed in an electric field will migrate at the same rate towards the + end of the field. When the DNA fragments are forced to migrate through a matrix like agarose or polyacrylamide gel, the gel acts like a sieve that the fragments must move through. Small fragments move easily through the pores of the gel, while the large fragments must snake or wind through the pores. Electrophoresis through the gel matrix causes the DNA fragments to separate by size with small fragments migrating faster than larger fragments.

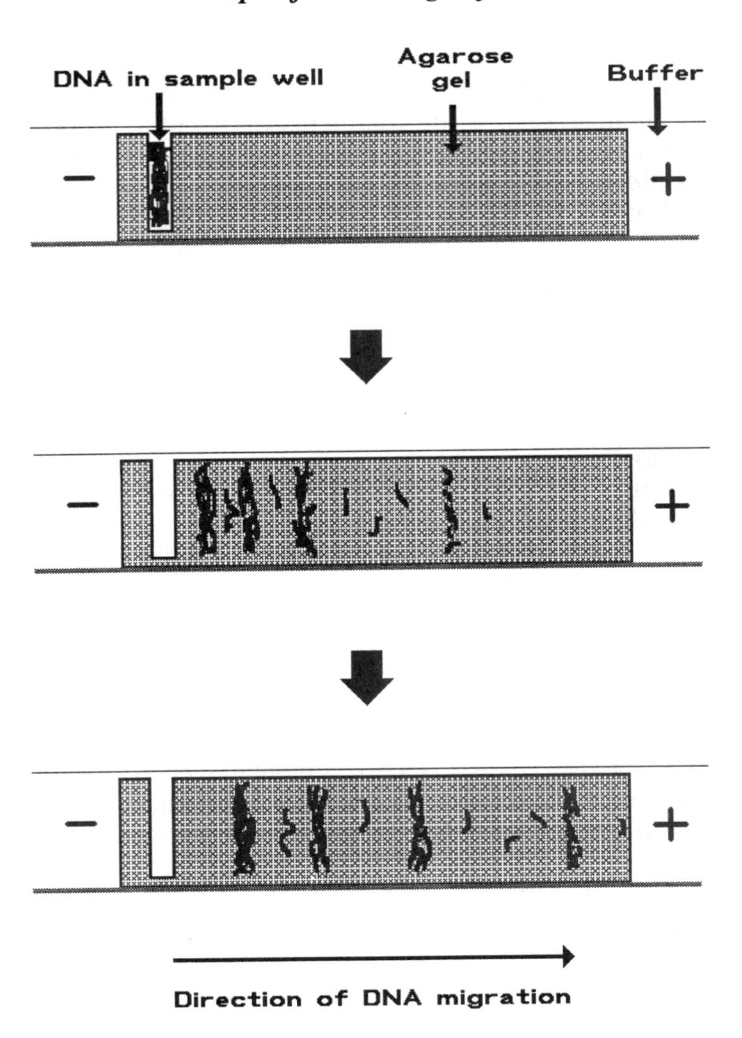

Figure 4.5. During agarose gel electrophoresis a DNA sample is placed in a well at one end of a slab of agarose gel submerged in a buffer and a current is applied to the gel. The negatively charged DNA fragments will migrate towards the positive end of the gel. As the DNA sample moves through the agarose matrix, the different fragments will separate according to size. The specific DNA fragments generated by restriction enzyme digestion will form bands in the gel while the less frequently occurring non-specific fragments will occur at a lower level throughout the gel. Staining will reveal the intense bands caused by specific restriction fragments but the non-specific fragments are generally present at too low a level for detection.

for characterizing DNA fragments of different size ranges. Fragments of less than 1,000 base pairs are best sized on polyacrylamide gels, while fragments of 1,000 to 25,000 base pairs are generally examined on agarose gels. With special electrophoretic techniques, such as pulsed-field electrophoresis, fragments of several hundred thousand base pairs can be resolved by agarose gel electrophoresis. For certain techniques, such as determination of the nucleotide sequence of a DNA fragment, it may be necessary to denature the double-stranded DNA to single-stranded fragments. A denaturing agent such as sodium hydroxide, urea, or formamide is included in the gel to denature the DNA and ensure that fragments run as single strands during electrophoresis.

Mobility of DNA fragments during electrophoresis is inversely dependent on fragment size, but the relationship between size and mobility is non-linear. A 4,000 base pair fragment does not migrate twice as fast as an 8,000 base pair fragment and one-half as fast as a 2,000 base pair fragment. The logarithm of fragment size plotted versus distance migrated approximates a straight line over a portion of the graph (Figure 4.6). Although this graph becomes non-linear for larger DNA fragments (greater than 8,000 base pairs in size), the size of a DNA fragment can be estimated by comparison of the migration of the fragment with the relative migration of DNA fragments of known sizes, generally called size standards. The size of the unknown fragment can be estimated by plotting the logarithm of the sizes of the known fragments versus the distance each fragment migrated to establish a standard curve. The distance that the unknown fragment migrated is measured and plotted on the standard curve, then the size of the fragment estimated by comparison with the sizes of the standards.

When a sample of chromosomal DNA that has been digested with a restriction enzyme is loaded in the sample well of an agarose gel and electric current applied to the gel, the DNA fragments migrate into the gel and separate by size. When many fragments of the same size are present in the sample (such as the specific fragments generated by the restriction enzyme), these fragments migrate together and form a band in the gel (Figure 4.5). The non-specific

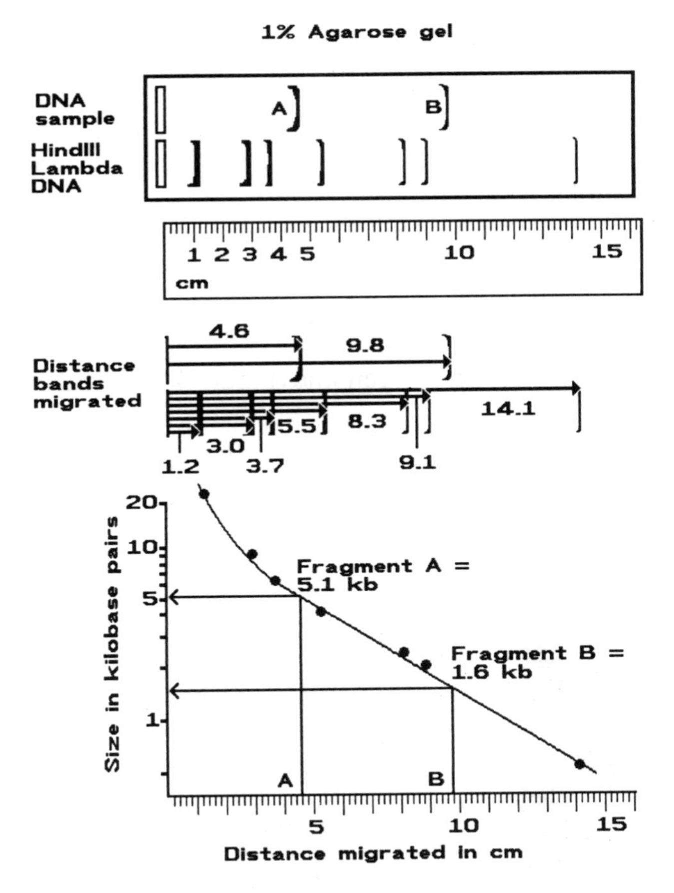

Figure 4.6. Use of DNA reference standards to determine the size of DNA fragments. A DNA sample containing fragments of undetermined size can be electrophoresed in an agarose gel along with a DNA sample that contains fragments of known size to be used as size references. Bacteriophage lambda DNA digested with restriction enzyme *Hin*dIII, for example, is a frequently used size reference that contains DNA fragments of 23,130, 9,416, 6,682, 4,361, 2,322, 2,027, and 564 base pairs as determined by nucleotide sequence analysis. To determine the sizes of fragments A and B, the distance traveled by the size references and by each fragment is measured. For each reference fragment, the log of the fragment size is plotted against the distance migrated by that reference fragment. A line drawn through the reference standards then can be used as a standard curve to determine the size of fragments A and B. Measurement indicates that fragment A traveled 4.6 cm, which can be plotted on the standard curve to obtain a size estimate of 5,100 base pairs.

DNA fragments that have one restriction enzyme end and one sheared end also migrate in the gel according to their sizes and form a background of DNA fragments. When the gel is stained and visualized, only size classes of DNA fragments with at least 1-2 nanograms of DNA in a band on the gel can be seen. The specific restriction fragments are seen as discrete bands on the gel, while the non-specific fragments are generally present at too low a level to be detected.

The number of bands generated by digestion of a DNA sample with a restriction enzyme is dependent on the complexity or size of the DNA genome (Figure 4.7). A small

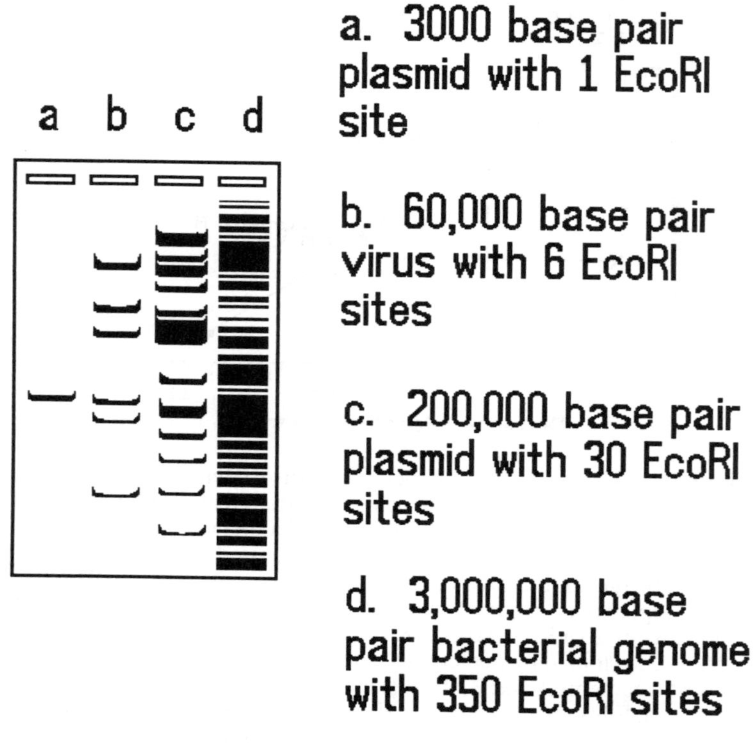

Figure 4.7. Increase in genome size is accompanied by an increase in the number of restriction fragments generated by digestion of the chromosomal DNA. The ability to resolve fragments of similar size decreases as the fragment number increases and regions containing many similar fragments may blur into broad, indistinct bands.

Agarose gel of digested DNA

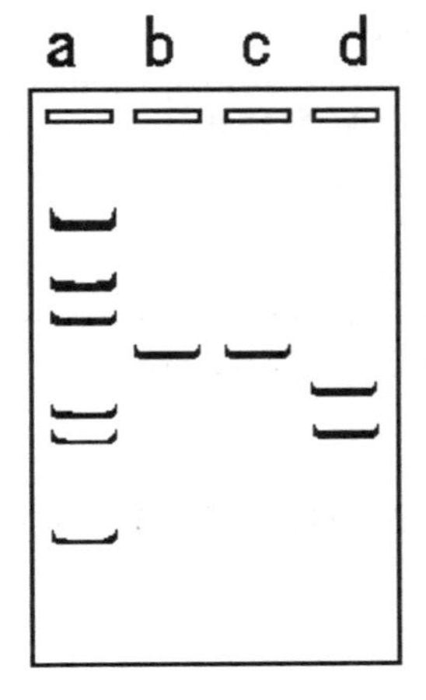

a. HindIII lambda DNA standards
b. EcoRI digested plasmid DNA, 5.1 kb
c. BamHI digested plasmid DNA, 5.1 kb
d. EcoRI + BamHI double-digested plasmid DNA, 3.0 + 2.1 kb

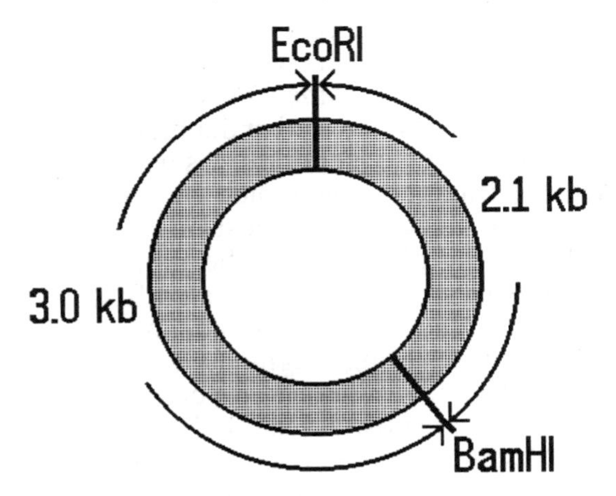

Figure 4.8. Construction of a restriction endonuclease map. Sequential digestion of a DNA with more than one enzyme can be used to position the cleavage sites relative to one another. In the example above a circular plasmid DNA contains a single *Eco*RI site (shown in lane b) and a single *Bam*HI site (shown in lane c). Digestion of the DNA with both enzymes (shown in lane d) reveals that these sites are located 2,100 base pairs from each other. Unlike a genetic map, a restriction map directly reflects the physical properties of a DNA molecule because the restriction map is a direct result of the nucleotide sequence of the molecule.

circular DNA molecule, such as a 3,000 base pair plasmid, might contain a single cleavage site for a restriction enzyme and form a unique band on an agarose gel. As the genome size increases, more fragments result from digestion with the same restriction enzyme. A 50,000 base pair bacteriophage DNA, like bacteriophage lambda, might be cleaved to generate 6 specific fragments, a 5 x 10^6 base pair bacterial genomic DNA might yield 500 fragments, and a 5 x 10^9 base pair mammalian DNA sample might yield several thousand specific DNA fragments.

Specificity of restriction fragment patterns

Because restriction enzymes cleave at specific sites determined by the nucleotide sequence of the DNA fragment being digested, restriction enzymes that recognize and cleave at different sequences will generate different fragment patterns from the same DNA sample. A DNA sample can be successively digested or double-digested with more than one enzyme to generate a set of fragments different than the fragments generated by either enzyme alone (Figure 4.8). Successive digestion with different enzymes is the principle used in the generation of a restriction map. A DNA sample is digested first with one restriction enzyme, then with a second, then with both together, and all the resulting fragments are sized by gel electrophoresis. Sizes of fragments are calculated and all of the restriction sites are positioned relative to one another on the DNA molecule to generate a restriction map.

In contrast to a genetic map, which is based on recombinational frequency and provides a relative positioning of genetic markers on a chromosome, a restriction map consists of a relative positioning of restriction cleavage sites on a DNA molecule and is a direct representation of some of the nucleotide sequence features of the chromosome. When a genetic map is combined with a restriction map, it is possible to position genes on specific DNA fragments of the chromosome (Figure 4.9). Restriction enzymes provide a tool that allows DNA fragments to be specifically aligned, allowing the construction of restriction maps that establish specific reference points on DNA molecules.

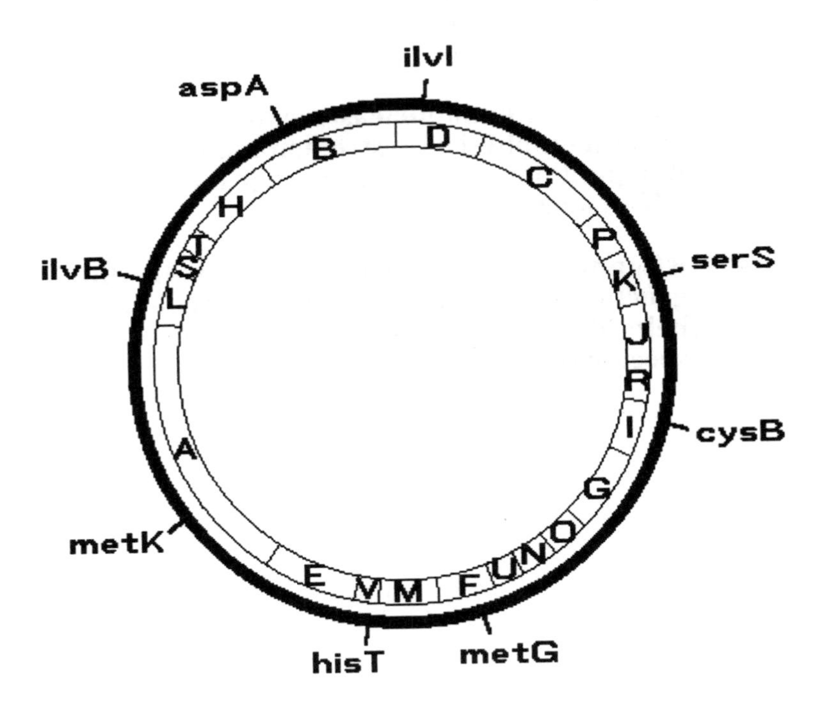

Figure 4.9. Alignment of a genetic map with a restriction map. The genetic map of eight genes involved in amino acid metabolism is shown aligned with the restriction map of the *Not*I cleavage sites present in the chromosomal DNA. While the genetic map uses units of transfer time or recombinational frequency, the restriction map is based on the nucleotide sequence of the chromosomal DNA. The his*T* gene, for example, can be demonstrated to occur within *Not*I restriction fragment V. Alignment of a genetic map with a restriction map establishes a relationship between gene location and specific DNA fragments.

Summary

1. A genetic map allows the ordering of genes on a DNA molecule but the order of the genes does not necessarily correlate precisely with the physical distance between genes.

2. Restriction endonucleases cleave DNA in a site-specific manner and allow DNA fragments of random sizes to be converted to specific fragments.

3. Restriction endonucleases can be used to generate a physical restriction map of DNA that positions restriction cleavage sites on the DNA molecule.

4. Alignment of a restriction map with the genetic map of the same DNA molecule can help identify DNA fragments that contain specific genes.

5

Making New DNA Molecules

"Potter traced down into the coiled-coil helices of the DNA chains with a dawning wonder. The altered gene pattern held true, but it was a pattern, Potter realized, which hadn't been seen in humankind since the second century of gene-shaping." — The Eyes of Heisenberg, Frank Herbert, 1966

The basic principles of genetic engineering have for centuries been applied to the breeding of plants and animals for specific, desired phenotypic traits: breed two plants or animals to facilitate recombination of genes, search through the offspring for the desired combination of phenotypic traits, then repeat the breeding process with the "improved" progeny. Disease resistance, milk production, wool characteristics, meat yield, and size of seed kernel are all genetic traits that have been specifically selected using conventional genetic breeding methods. While these conventional genetic approaches have been extremely successful in the development of agriculturally significant plant and animal strains, the methods can be very slow and remarkably inefficient. During normal sexual reproduction, the genetic traits of the parents mix in a fairly random manner and obtaining the desired combination of phenotypic characteristics may require searching through thousands of progeny. At times, the appearance of the desired trait may depend on the occurrence of a spontaneous change in DNA sequence, a mutation, that can occur at extremely low frequencies. Mutation rates of 1 in 10^6 copies of a gene are not unusual, and finding a mutation that occurs at this frequency might require examining millions of individuals looking for the desired phenotype. Conventional genetic methods are also limited by existing genetic barriers that make transfer of a desired trait from one species to another extremely difficult, if at all possible.

As early as the mid-1960's, long before the advent of recombinant DNA methods, science fiction writers had begun to speculate that advances in genetic technology would allow the specific manipulation of DNA to generate novel, desired genetic arrangements. The discovery of restriction enzymes that cleave DNA in a site-specific manner to generate discrete gene fragments might be considered the first crucial element in the development of the ability to make novel recombinant DNA molecules *in vitro*, or in a test tube. The second crucial contribution to recombinant DNA technology was the development of the ability to seal the breaks made in DNA by restriction enzymes and make new arrangements of DNA fragments.

Restriction enzymes allow recombination in a test tube

Shortly after the discovery that restriction enzymes can be used to cleave DNA into fragments with specific, defined ends, researchers realized that the short, cohesive termini generated by many restriction enzymes could be used to make new combinations of DNA fragments - recombinant DNA molecules. As discussed previously, single-stranded DNA can anneal to a complementary sequence, using the A-T, G-C base-pairing principles, to form double-stranded DNA. When a DNA molecule is digested with a restriction enzyme that generates short cohesive termini, the single-stranded ends of the resulting DNA fragments can anneal to reform the original DNA molecule or can anneal with the cohesive ends of other DNA fragments to form new DNA fragment combinations (Figure 5.1).

While normal recombination processes that give rise to new genetic arrangements usually require fairly high amounts of sequence similarity between the DNA fragments undergoing recombination (Figure 5.2), restriction enzymes can be used to facilitate "site-specific, illegitimate recombination". This process allows the recombination of

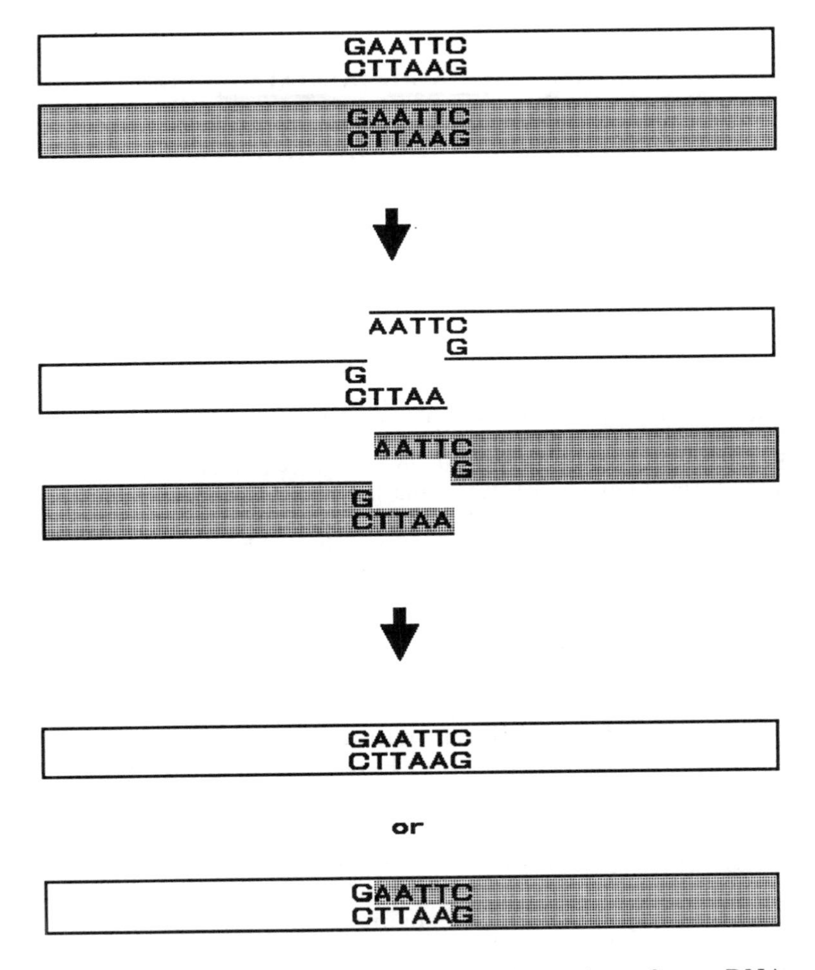

Figure 5.1. Restriction enzymes allow formation of new DNA molecules. When a DNA fragment is cleaved with a restriction enzyme like *Eco*RI (recognition site GAATTC), the resulting short cohesive ends (AATT) can anneal to reform the original DNA fragment or can anneal with the ends of different DNA fragments to form new combinations of DNA fragments. These new arrangements of fragments are called recombinant DNA molecules.

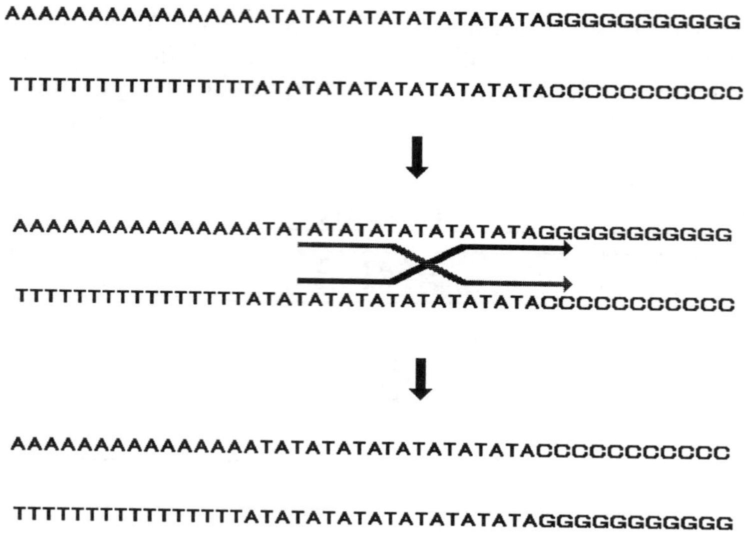

AAAAAAAAAAAAAAAATATATATATATATATATAGGGGGGGGGG

TTTTTTTTTTTTTTTTTATATATATATATATATATACCCCCCCCCCC

↓

AAAAAAAAAAAAAAAATATATATATATATATATAGGGGGGGGGG

TTTTTTTTTTTTTTTTTATATATATATATATATATACCCCCCCCCCC

↓

AAAAAAAAAAAAAAAATATATATATATATATATACCCCCCCCCCC

TTTTTTTTTTTTTTTTTATATATATATATATATATAGGGGGGGGGG

Figure 5.2. Normal DNA recombination processes generally occur between very similar or homologous regions of DNA molecules. Recombination would be unlikely to occur between the stretch of A's and the stretch of T's or between the stretch of G's and the stretch of C's in the two DNA sequences above. Recombination could occur between the two TA-rich sequences as indicated by the two arrows, generating the two new recombinant DNA molecules shown below.

DNA fragments that are similar **only** in the presence of a restriction enzyme cleavage site. In principle, two genes that are completely different will rarely recombine during natural biological processes. However, in the test tube, restriction enzymes can be used to cleave the two unrelated DNA molecules to generate short, cohesive single-stranded ends and the ends allowed to anneal to generate new combinations of DNA fragments.

The initial attempts at using restriction enzymes to generate new DNA molecules were not very sophisticated. Two DNAs were digested with the same restriction enzyme, mixed together, and inserted into bacteria. The resealing of the nicks in the DNA molecules was allowed to take place in the bacterial cells. Needless to say, this was an extremely inefficient process and obtaining the desired recombinants was difficult.

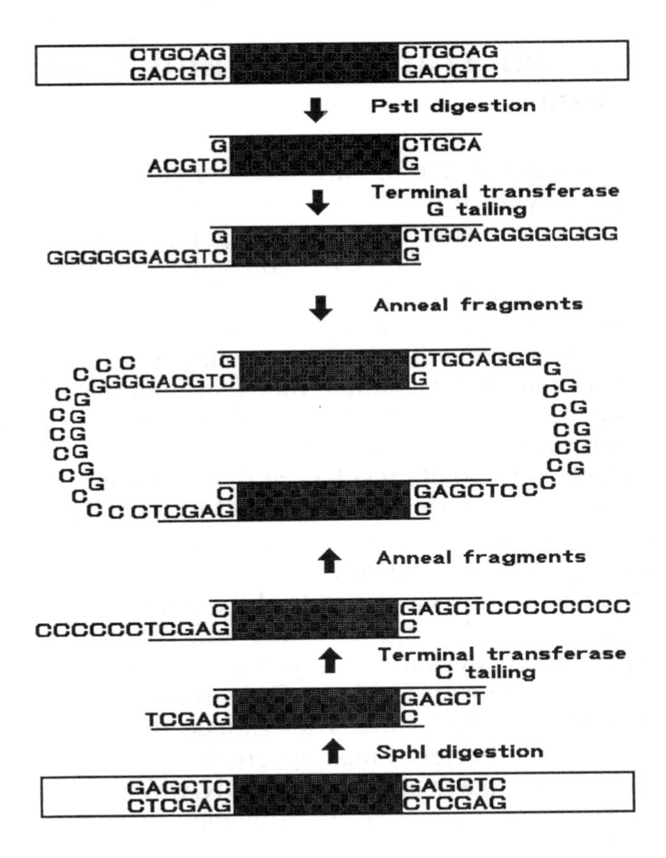

Figure 5.3. Use of the enzyme terminal transferase to form recombinant DNA molecules. A restriction enzyme that produces a 3' cohesive terminus (such as *Pst*I producing the end CTGCA) is used to generate the first DNA fragment. Terminal transferase and dGTP are then used to add a stretch of dG residues to the 3' ends of the fragment, restoring the *Pst*I site (CTGCAG). A second DNA fragment generated by digestion with a different restriction enzyme (such as *Sph*I producing the end GAGCT) is treated with terminal transferase and dCTP to add a stretch of dC residues to the ends of the fragment, restoring the *Sph*I site (GAGCTC). When the two fragments are mixed together, the poly-C tails on the ends of the first DNA fragment will anneal, or form base pairs, with the poly-G tails on the ends of the second DNA fragment. The resulting annealed circle will contain two single-stranded gaps that can be repaired to double-stranded DNA when the molecule is inserted into a cell. Note that because the poly-G tails can **only** form base pairs with poly-C tails, circularization of a DNA fragment, or annealing of one end of a DNA fragment to the other end of the fragment, cannot occur.

Terminal transferase and "tailing" reactions

The efficiency of the *in vitro* recombination process was increased by the use of the enzyme terminal transferase, a protein that can add deoxynucleotide residues onto the 3' hydroxyl terminus of a DNA molecule. This approach to resealing the nicks in the DNA molecule, referred to as terminal transferase "tailing", usually involves digesting one DNA sample with a restriction enzyme that leaves a 3' terminal cohesive terminus, like that generated by *Pst*I (Figure 5.3). The 3' ends of this DNA fragment are treated with terminal transferase in the presence of dGTP to add a stretch of 25 to 50 dG residues at the 3' termini. Note that this dG tailing reaction restores the G residue that was lost from the *Pst*I site during the cleavage reaction. The second DNA fragment is tailed with dCTP to produce a stretch of 25 to 50 dC residues at the ends of the fragment. When the two tailed DNA fragments are mixed together, the complementary dG and dC tails efficiently anneal with one another to form a recombinant DNA molecule containing two single-stranded gaps. This gapped molecule can be inserted into a bacterial host, where the gaps are repaired by the cell to make double-stranded DNA. While this tailing process greatly increases the efficiency of generation of recombinants, it adds stretches of GC residues onto the ends of the manipulated DNA fragments, which can be undesirable.

DNA ligase

The capability to seal nicks *in vitro* was greatly enhanced by the discovery of enzymes called DNA ligases. Like restriction enzymes and terminal transferase, these are naturally occurring proteins that play an important role in the normal metabolism of DNA in the cell. Ligase requires an additional co-factor compound that participates in the joining of a 5' phosphate to a 3' hydroxyl at a nick in a DNA molecule (Figure 5.4). The ligase purified from *E.*

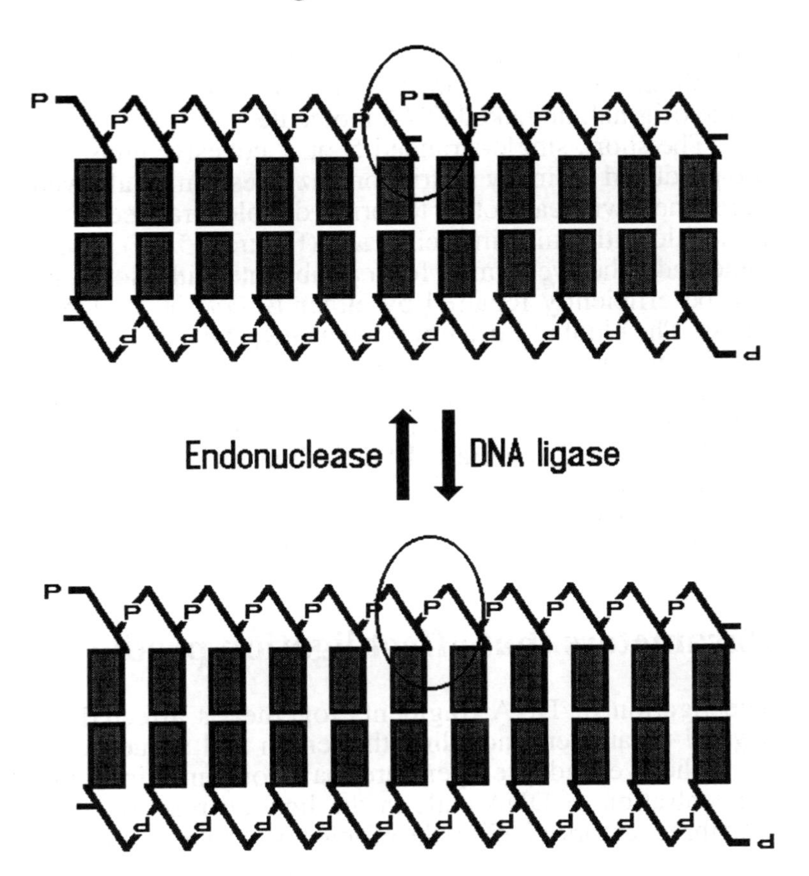

Figure 5.4. Role of DNA ligase in DNA metabolism. DNA ligase can join the 3' hydroxyl residue of one sugar residue to the 5' phosphate of the adjacent sugar residue at a nick in the sugar-phosphate backbone of DNA, indicated in the circle. This reaction is the reverse of the action of an endonuclease, which introduces a nick between adjacent sugar residues.

coli uses the compound NAD^+ as a co-factor, while T4 DNA ligase, an enzyme that is made from a gene present in the bacteriophage T4, uses ATP as a co-factor. Ligase uses the co-factor to activate the 5' phosphate of the nick being sealed and the NAD^+ is broken down to NMN or the ATP to AMP during the nick-sealing reaction. The resultant 5' to 3' phosphodiester bond is indistinguishable from that present in the original, unnicked DNA molecule.

The short, single-stranded 5' or 3' cohesive ends that are produced by many restriction enzymes can be allowed to reanneal with each other to form a double-stranded DNA molecule with a nick in each strand (Figure 5.5 b,c). These annealed cohesive termini form a substrate with nicks that can be efficiently repaired by either *E. coli* or T4 DNA ligase. The blunt or flush termini that are produced by some restriction enzymes, such as *Sma*I, cannot anneal to form the nicked double-stranded DNA substrate required by *E. coli* ligase and blunt ended termini cannot be sealed together by this enzyme. In contrast, T4 DNA ligase can ligate blunt ends together (Figure 5.5 d). Because of its ability to ligate either blunt or cohesive ends, T4 DNA ligase is most frequently used for resealing DNA fragments in a test tube.

Parameters that affect ligation products

The ligation of DNA fragments together is affected by several parameters including the length and sequence of the cohesive ends, temperature, salt concentration, and concentration of DNA ends in the ligation reaction. The first three factors all affect the efficiency or rate with which the cohesive ends anneal to form the nicked DNA substrate and can be somewhat minimized by using standardized conditions for all ligation reactions. The concentration of DNA ends in the reaction mix, however, actually affects the type of DNA product that ligation generates.

A ligation reaction typically contains several different DNA fragments generated by digestion with a single restriction enzyme. Since the cohesive ends on the fragments are identical and complementary to each other, an individual cohesive end can anneal with any other end in the mixture of DNA fragments. In a ligation mix with only three different types of DNA fragment A, B, and C generated by digestion with *Eco*RI (Figure 5.6), all fragments have the 5' cohesive terminus AATT and a fragment A end could anneal with any other end. Annealing

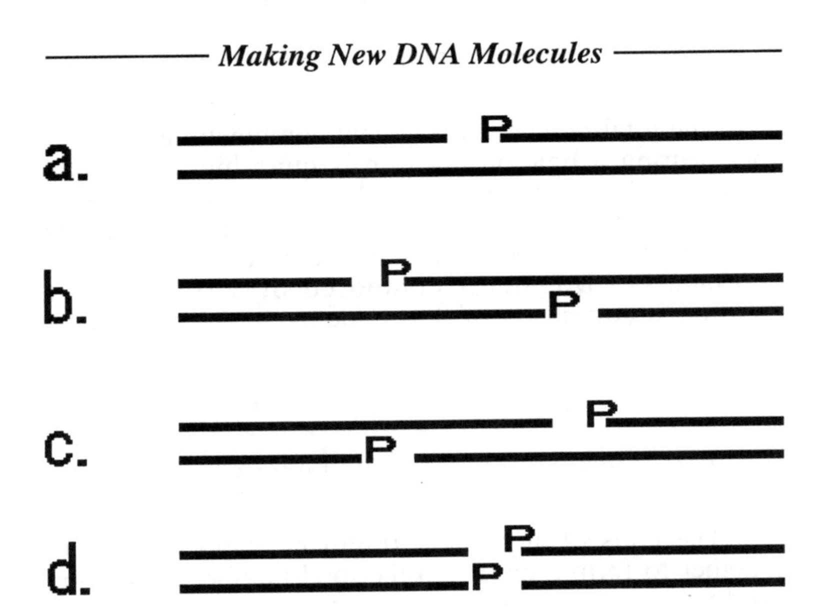

Figure 5.5. Nicks that can serve as a substrate for DNA ligase must possess a 3' hydroxyl and an adjacent 5' phosphate residue. Both *E. coli* and T4 DNA ligase can seal a) a single nick in one strand of a double stranded DNA, b) opposing nicks generated by a restriction enzyme that leaves a 5' cohesive terminus, or c) opposing nicks generated by a restriction enzyme that leaves a 3' cohesive terminus. T4 DNA ligase, but **not** *E. coli* DNA ligase, can also seal d) the ends of adjacent blunt-end fragments.

with a different fragment in a bimolecular reaction would generate new fragment combinations (AA, AB, or AC). This annealing is not limited to two fragments, but can involve many different fragments annealing with each other to produce different linear fragment combinations (AAB, ABB, ABC, AABBC, CABA, etc). Any particular end could also anneal with the other end of the same DNA molecule in an intramolecular reaction to form a circle. As the cohesive ends in a mixture of DNA fragments anneal, a complex mixture of both linear and circular fragments combinations begins to form. When DNA ligase is added to such a complex annealing mixture of fragments, the nicks in the annealed cohesive ends are repaired and the arrangements of fragments become permanent.

The total number of fragment combinations that can form during a ligation reaction is quite high. The total number of linear fragment possibilities can be **estimated** by summing the number of possible ligation products composed of 1 fragment + products composed of 2 fragments + fragments composed of 3 fragments + fragments composed of 4 fragments + ... This can be represented by the equation:

$$\text{Number of linear products} = N + N\times N + N\times N\times N + N\times N\times N\times N + \ldots$$
$$= N + N^2 + N^3 + N^4 + \ldots$$

where N equals the number of different fragments in the ligation reaction.

The ends of any linear fragment could also ligate together to form a circle, so the total number of ligation products must include both linear and circular ligation products. Since each linear ligation product could also be ligated into a circular ligation product, the number of potential circular products must be at least as large as the number of potential linear ligation products. An estimate of the minimum number of potential ligation products would be indicated by:

$$\text{Number of total products} = 2[N + N^2 + N^3 + N^4 + \ldots]$$

where N equals the number of different fragments in the ligation reaction.

For the 3-fragment ligation shown in Figure 5.6, the total number of ligation products could be estimated as $2[3 + 3^2 + 3^3 + 3^4 + \ldots]$. If all products could form with equal frequency, it is obvious that the total number of different ligation products would be extremely high. Fortunately, not all ligation products form with the same probability.

The nature of the ligation products that result from ligating particular fragments together can be approximated by a mathematical formula that takes into account the relative ability of a DNA fragment to anneal with its own end or that of a different fragment. This formula is, at best, an approximation of the products of a ligation reaction and can be simplified to the following principle: in a concentrated ligation reaction, fragments are more likely

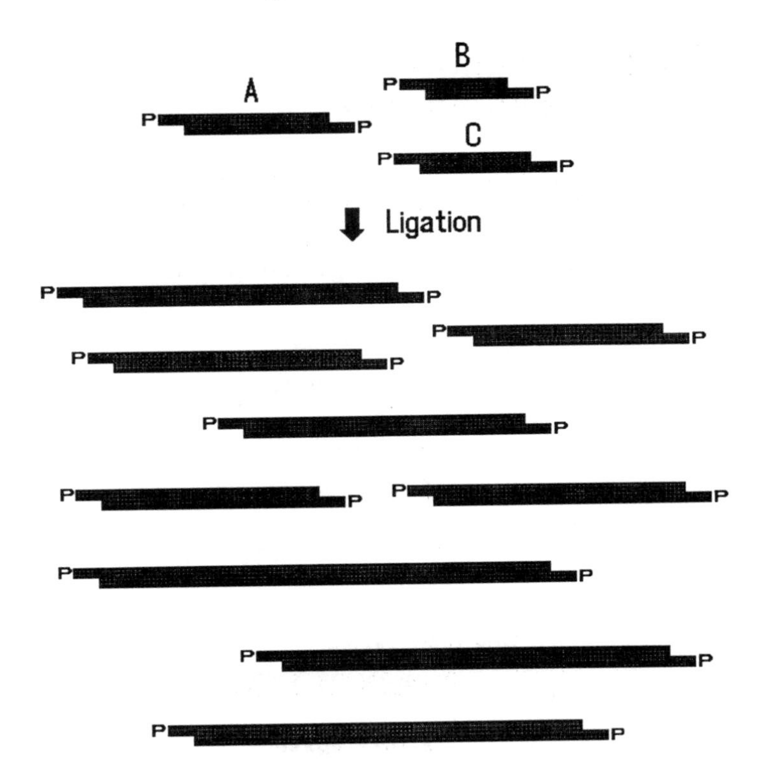

Figure 5.6. Different fragment combinations that result on ligation of a mixture of three different DNA fragments generated by digestion with *Eco*RI. In addition to simple dimers (AA, BB, AB, etc.), multimers containing several DNA fragments can form (ACA, CAB, BCA, etc.). In addition to the linear forms shown, circular molecules form when ends of the same DNA molecule anneal together and ligate with one another.

to ligate to other fragments to make linear combinations of fragments called multimers, while in a dilute reaction, fragments are more likely to ligate to their own ends to form circles. Although ligation conditions can be manipulated to **favor** particular types of products, it is difficult to force a ligation reaction to give a specific desired fragment combination. Ligation generally gives mixtures of many different linear and circular combinations of DNA fragments.

Inhibition of undesired ligation products with DNA phosphatase

It is at times possible to reduce the undesired ligation products. A DNA fragment can be treated with the enzyme DNA phosphatase, for example, to remove the 5' phosphate residue that is required for ligation (Figure 5.7). The

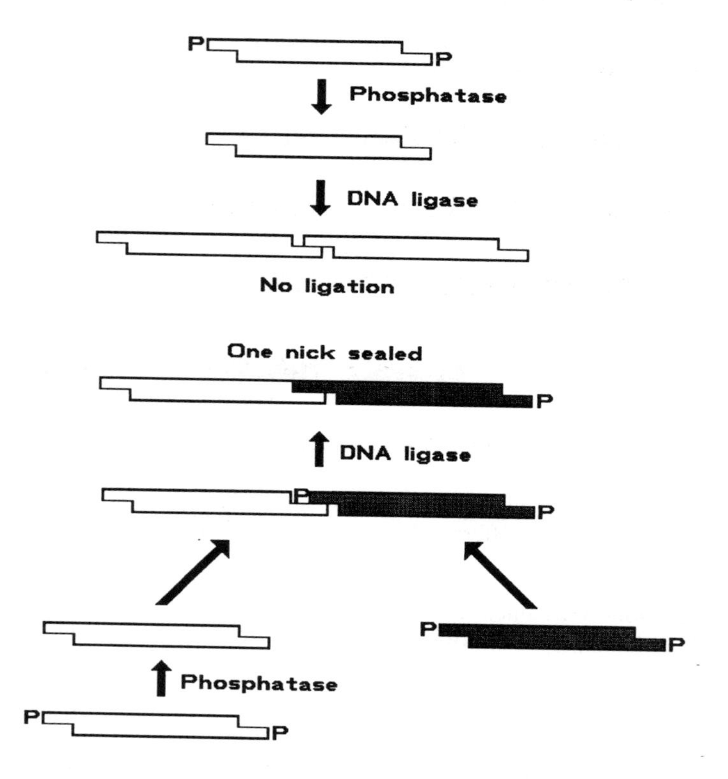

Figure 5.7. DNA phosphatase removes the 5' phosphate from DNA fragments. Phosphatase-treated DNA fragments can anneal with one another but the nicks cannot be sealed by DNA ligase, which requires the presence of a 5' phosphate (top portion of illustration). A fragment that has been treated with phosphatase can anneal with a second DNA fragment that retains a 5' phosphate (bottom portion of illustration). Treatment with DNA ligase will seal one of these two nicks to generate a recombinant DNA ligation product. This strategy is often used to prevent circularization of a DNA fragment and force ligation to a second fragment to produce a desired recombinant.

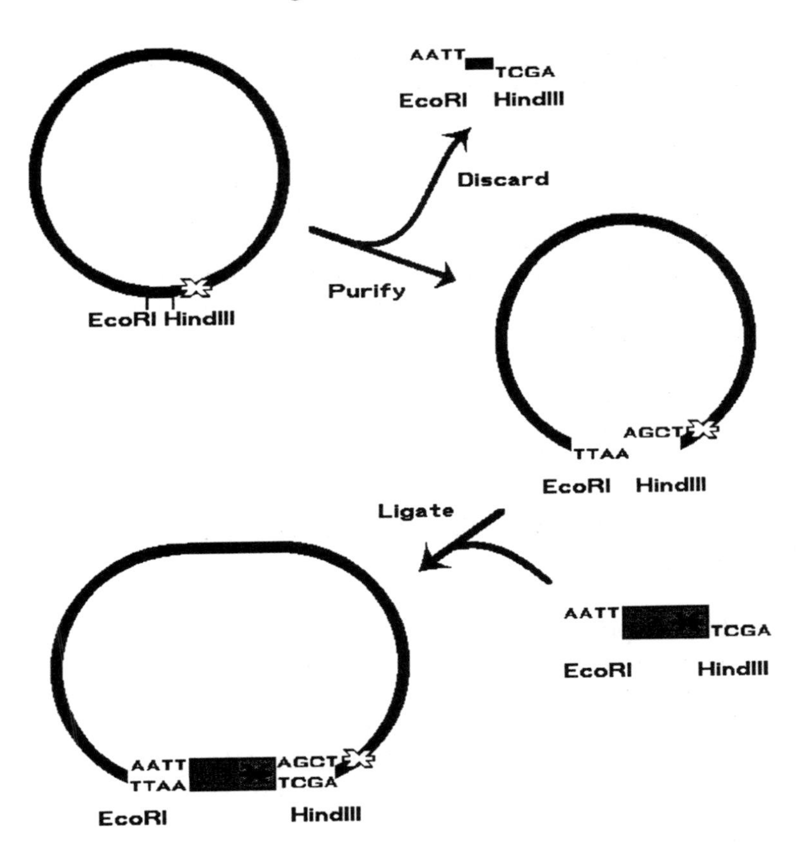

Figure 5.8. Digestion with two different restriction enzymes to obtain a desired ligation product. The short cohesive ends generated by restriction enzymes will only anneal with ends of the same sequence. A circular DNA that contains a single *Eco*RI site (GAATTC) and a single *Hin*dIII site (AAGCTT) can be digested with both enzymes to generate two different DNA fragments with AATT at one end and AGCT at the other end. The large fragment can be purified away from the smaller fragment and mixed with a second DNA fragment also generated by digestion with *Eco*RI and *Hin*dIII and having AATT at on end and AGCT at the other end. Since the AATT end cannot anneal to the AGCT end, the two DNA fragments cannot circularize by themselves. The two complementary ends will anneal to form a large circle consisting of one large and one small DNA fragment. Note that because the cohesive ends must anneal with their complementary sequences (DNA-AATT-3' with 3'-TTAA-DNA and DNAAGCT-3' with 3'-TCGA-DNA), only one orientation of the fragments is possible and the ends marked with * must always ligate together.

phosphatase-treated end can no longer participate in a ligation reaction, so the ends of the fragment are no longer capable of ligating with one another to form a circular molecule. If a second DNA fragment that has not been treated with phosphatase is allowed to anneal with the first fragment, the 5' phosphate residues on the second fragment allow the two fragments to ligate to form a circle containing two nicks where the phosphatase treatment removed the 5' phosphates. This nicked circle can be inserted into a host cell, where the missing phosphates are replaced and the nicks are sealed to make an intact circular DNA. The use of phosphatase prevents the ligation of a fragment to itself or another fragment of the same type.

Ligation of fragments in forced orientations

Just as it is possible to use phosphatase to affect the outcome of ligation reactions, digestion with more than one restriction enzyme can be used to favor or force the occurrence of desired ligation products (Figure 5.8). A DNA fragment that has been digested with *Eco*RI to generate the 5' cohesive end AATT at one end and with *Hin*dIII to generate the 5' cohesive end AGCT at the other end will not be able to ligate and form a circle with a DNA fragment that contains either two *Eco*RI or two *Hin*dIII ends. However, a DNA fragment with both an *Eco*RI and a *Hin*dIII cohesive terminus will efficiently ligate to the first fragment to form a circle. The ability of cohesive termini to anneal only with a complementary sequence allows selective ligation of certain fragment orientations or combinations.

Ligation of ends to dissimilar termini

The selective annealing of cohesive termini to other termini with the same nucleotide sequence would appear to put limits on the versatility of using ligase to rearrange DNA fragments. This apparent problem can be overcome by using the blunt-end ligating ability of T4 ligase. Because blunt ends do not have to anneal with one another to allow ligation to proceed, blunt ends generated by restriction enzymes that recognize and cleave at two different sequences can be efficiently joined together by T4 DNA ligase. For example, *Sma*I cleaves the sequence CCCGGG to generate a blunt terminus

```
–CCCGGG–      SmaI       –CCC     GGG–
–GGGCCC–   ---------->   –GGG     CCC–
```

and the enzyme *Hpa*I cleaves the sequence GTTAAC to generate a different blunt terminus

```
–GTTAAC–      HpaI       –GTT     AAC–
–CAATTG–   ---------->   –CAA     TTG–
```

These two ends can be ligated together with T4 DNA ligase to generate a new sequence

```
–CCC     AAC–    T4 DNA ligase     –CCCAAC–
–GGG     TTG–   --------------->   –GGGTTG–
```

that is a fusion of the *Sma*I and *Hpa*I sites and can no longer be cleaved by either of these enzymes.

The utility of the blunt-end ligation reaction becomes even more apparent with the realization that it is possible to convert either a 5' or a 3' cohesive end to a blunt end. This can be accomplished with other DNA modifying enzymes, including DNA polymerase and various nucleases. *E. coli* DNA polymerase I, an enzyme normally involved in DNA synthesis, recombination, and repair, has one 5' to 3' DNA synthesizing activity and two DNA degrading activities. The DNA synthesis reaction requires a double-stranded template that consists of a primer with a free 3' hydroxyl residue that is adjacent to a single-stranded region of DNA. DNA polymerase I uses the single-stranded region as a template and assembles the newly synthesized DNA strand from deoxynucleoside triphosphates (dNTP's):

```
NNNNNNN-OH                              + dNTP's
NNNNNNNNNNNNNNNNNNNNNNNNN

                               |  DNA polymerase I
                               v

NNNNNNnnnnnnnnnnnnnnnnnnnnnnnnnnnnn
NNNNNNNNNNNNNNNNNNNNNNNNNNNNNNNNNNN
```

where the newly synthesized DNA is indicated by lower case letters.

In the absence of dNTP's, DNA polymerase can also degrade DNA from a free 3' hydroxyl residue in a sort of reverse reaction of the DNA synthesis reaction that is called a 3' to 5' exonuclease activity:

```
NNNNNNNNNNNNNNNNNNNNNNNNNNNNNN-OH
NNNNNNNNNNNNNNNNNNNNNNNNNNNNNN

                               |  DNA polymerase I
                               v

NNNNNNNNNNNNNN-OH
NNNNNNNNNNNNNNNNNNNNNNNNNNNNNNNNNNNNNN
```

The other type of DNA degrading activity associated with DNA polymerase I is a 5' to 3' exonuclease activity that can remove nucleotides from the 5- phosphate end of a DNA molecule:

```
P-NNNNNNNNNNNNNNNNNNNNNNNNNNNNN
  NNNNNNNNNNNNNNNNNNNNNNNNNNNNN

                               |  DNA polymerase I
                               v

                         P-NNNNNNNNNNNNNN
NNNNNNNNNNNNNNNNNNNNNNNNNNNNNNNNNNNNN
```

This exonuclease activity can work in conjunction with the DNA polymerase activity to simultaneously remove nucleotides and replace them with newly synthesized DNA in a reaction known as “nick translation”:

```
NNNNNNNN     NNNNNNNNNNNNNNNNNNNNNNNNNNNNNNNNNNNNNN
NNNNNNNNNNNNNNNNNNNNNNNNNNNNNNNNNNNNNNNNNNNNNNNNNNNN
```

```
NNNNNNNNnnnnnnn     NNNNNNNNNNNNNNNNNNNNNNNNNNNNNNNN
NNNNNNNNNNNNNNNNNNNNNNNNNNNNNNNNNNNNNNNNNNNNNNNNNNNN
```

```
NNNNNNNNnnnnnnnnnnnnnnnnnnnnnnnn    NNNNNNNNNNNNNNN
NNNNNNNNNNNNNNNNNNNNNNNNNNNNNNNNNNNNNNNNNNNNNNNNNNNN
```

```
NNNNNNNNnnnnnnnnnnnnnnnnnnnnnnnnnnnnnnnnnnnnnn   NN
NNNNNNNNNNNNNNNNNNNNNNNNNNNNNNNNNNNNNNNNNNNNNNNNNNNN
```

The purified DNA polymerase I protein can be cleaved with a protease to generate a large polymerase fragment, called Klenow fragment, that retains the DNA synthesis and 3' to 5' exonuclease activities. Because Klenow fragment of DNA polymerase I can no longer degrade the 5' phosphate end of DNA, Klenow fragment is routinely used to convert a 5' cohesive terminus to a blunt end:

```
NNNNNNN          Klenow fragment       NNNNNNnnnn
NNNNNNNNNNN      --------------->      NNNNNNNNNN
```

Non-complementary, 5' cohesive ends generated by two different restriction enzymes can be converted to blunt ends with Klenow fragment of DNA polymerase I, and then ligated together with T4 DNA ligase. For example, *Eco*RI cleaves the sequence GAATTC to generate a 4-base cohesive terminus

```
—GAATTC—        EcoRI       —G       AATTC—
—CTTAAG—     ---------->    —CTTAA       G—
```

and the enzyme *Bam*HI cleaves the sequence GGATCC to generate a different 4-base cohesive terminus

```
—GGATCC—        BamHI        —G        GATCC—
—CCTAGG—      -------->      —CCTAG        G—
```

that cannot be efficiently ligated together. However, treatment with Klenow fragment of DNA polymerase I generates blunt ends that can be ligated to one another

```
—GAATT  GATCC—   T4 DNA ligase     —GAATTGATCC—
—CTTAA  CTAGG—  -------------->    —CTTAACTAGG—
```

to generate a fusion of the *Eco*RI and *Bam*HI sites that can no longer be cleaved by either of these enzymes. Klenow fragment can be used to convert any 5' cohesive terminus to a blunt end that can be efficiently ligated by T4 DNA ligase.

Many restriction enzymes generate 3' cohesive termini. All known DNA polymerizing enzymes synthesize DNA in a 5' to 3' direction and cannot use a 3' cohesive terminus as a substrate for DNA synthesis and these cohesive termini cannot be converted to blunt ends by DNA synthesis. Other DNA-degrading enzymes called exonucleases can be used to remove the short single-stranded extension of DNA to convert the 3' cohesive terminus to a blunt end. S1 nuclease is an exonuclease that is specific for single-stranded nucleic acids and will degrade single-stranded DNA or RNA while leaving double-stranded regions relatively intact. This enzyme is frequently used to convert 3' cohesive termini to blunt ends

```
—NNNNNN      N—    S1 nuclease    —N        N—
—N       NNNNN—   ------------>   —N        N—
```

T4 DNA ligase used in conjunction with other DNA modifying enzymes like Klenow fragment of DNA polymerase I and S1 nuclease allows the ligation of DNA fragments with ends generated by digestion with different restriction enzymes. These procedures have the disadvantage that the end-filling and joining process generally alters the nucleotide sequences so that the ligation product can no longer be cleaved with the same restriction enzymes to recover the two fragments. This disadvantage can be overcome by the use of short synthetic DNA fragments called oligonucleotide linkers or adaptors.

Linkers are generally short, chemically synthesized, double-stranded DNA fragments that contain cleavage sites for one or more restriction enzymes, while adaptors generally contain one blunt and one cohesive terminus. These short DNA fragments can be ligated onto a blunt terminus using T4 DNA ligase to add a convenient restriction enzyme cleavage site. When adding linkers, the modified end must be digested with the appropriate restriction enzyme to free the cohesive terminus for subsequent manipulation. These linkers allow a specific, desired restriction cleavage site to be added onto the end of a blunt terminus to modify the sites present on the ends of a DNA fragment (Figure 5.9).

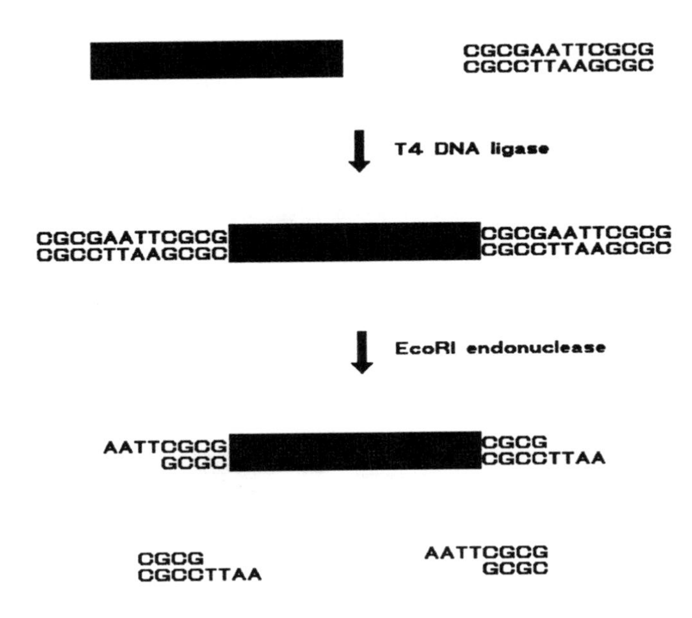

Figure 5.9. Addition of oligonucleotide linkers onto a DNA fragment. The linkers shown contain an *Eco*RI cleavage site (GAATTC) in the center of the 12-base sequence CGCGAATTCGCG. T4 DNA ligase can be used to ligate the linkers onto a blunt-end DNA fragment that contains no other *Eco*RI cleavage sites. Digestion with *Eco*RI endonuclease will free the cohesive AATT ends of the added *Eco*RI sites and release short fragments of the linkers. This procedure adds *Eco*RI sites onto the ends of the blunt-end fragment and allows ligation of the fragment into an *Eco*RI site in another DNA molecule.

One problem associated with the use of oligonucleotide linkers is the need to digest the linker-DNA ligation product to free the cohesive end contained within the linker. Any cleavage sites within the DNA fragment will also be cleaved by the restriction enzyme and smaller DNA fragments will be generated, rather than the intact fragment (Figure 5.10). This problem can be avoided by the use of the DNA restriction methylase that specifically modifies the restriction enzyme recognition site to prevent cleavage by the restriction enzyme (see Chapter IV). In this strategy (Figure 5.11), a blunt ended DNA fragment can be treated

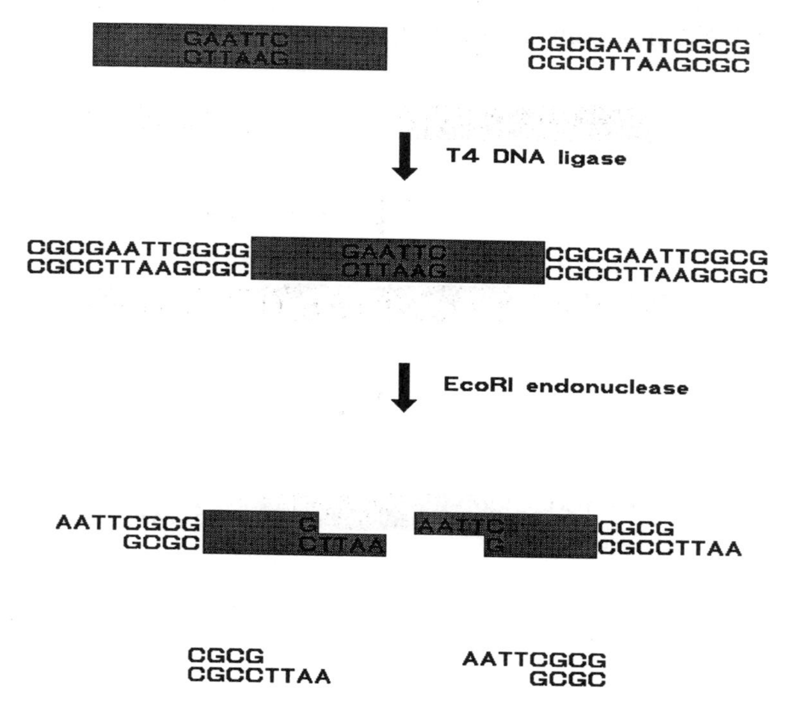

Figure 5.10. The presence of one or more internal cleavage sites for the enzyme used to free the linker ends will result in cleavage of the desired linked DNA fragment. Following the addition of the linkers by T4 DNA ligase, treatment with *Eco*RI will free the cohesive AATT ends of the added *Eco*RI sites, release short fragments of the linkers, and cleave at the internal *Eco*RI site (GAATTC). The original blunt-end DNA fragment has now been converted into two smaller DNA fragments with cohesive AATT ends.

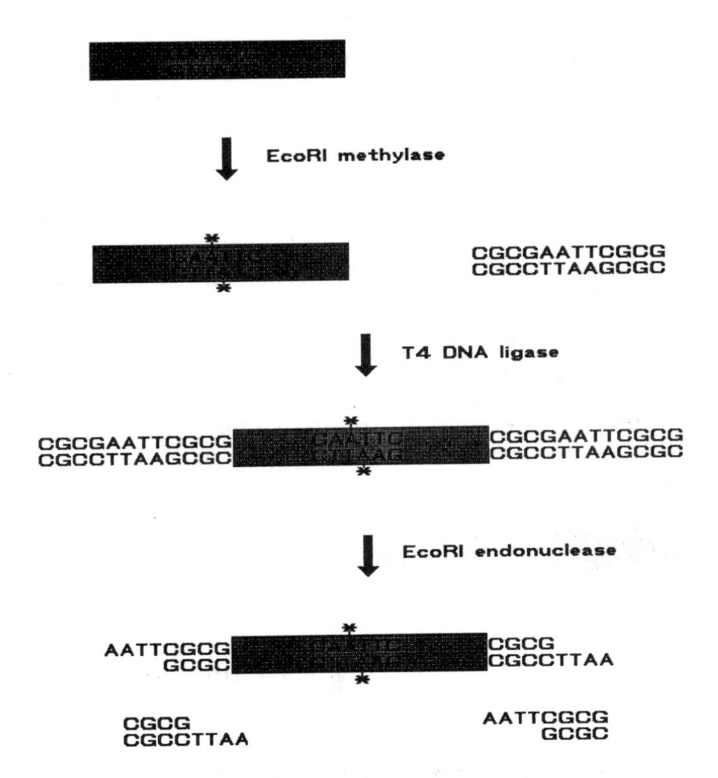

Figure 5.11. DNA methylase can modify internal restriction sites to prevent cleavage by a restriction enzyme. Methylation prevents cleavage of the internal sites during the digestion reaction used to free the cohesive ends of the linkers. The methylation reaction must be performed with a methylation enzyme that has the same recognition specificity as the restriction enzyme used to free the linkers. *Eco*RI methylase must be used to prevent cleavage at GAATTC by *Eco*RI, but will not prevent cleavage at GGATCC by *Bam*HI. Methylation of the internal sites in the DNA fragment must be done prior to the addition of linkers or the linkers will also be modified and protected from cleavage.

with restriction methylase to modify the fragment prior to the addition of the linkers. This methylation step will prevent cleavage of any of the restriction enzyme sites that might occur within the fragment. The linkers can then be ligated onto the modified fragment. During the subsequent digestion with restriction enzyme, the unmodified cleavage sites within the linkers will be digested while the modified DNA fragment will remain uncleaved and intact.

The problem of internal restriction enzyme cleavage sites can also be avoided by using oligonucleotide adaptors that contain the desired single-stranded cohesive terminus. This eliminates the need for restriction endonuclease digestion of the adaptor-DNA ligation product and facilitates the cloning of intact DNA fragments with the desired added termini (Figure 5.12).

The commercial availability of oligonucleotide linkers and adaptors, combined with T4 DNA ligase and other DNAmodifying enzymes, gives DNA researchers the capability to extensively modify DNA fragments. This capability contributed significantly to the development of

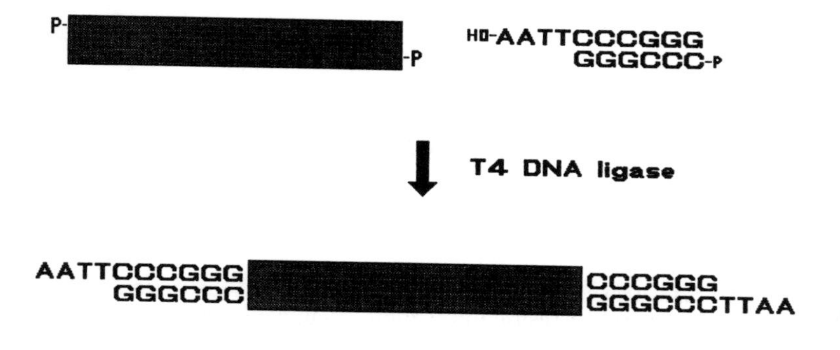

Figure 5.12. Addition of oligonucleotide adaptors onto a DNA fragment. Adaptors are composed of two different synthetic oligonucleotides of different length and sequence. When these two sequences are annealed together, they form a short DNA segment with one blunt end and one end corresponding to a cohesive terminus generated by a restriction enzyme. The adaptor shown is essentially an *Sma*I linker (CCCGGG) with the four-base cohesive *Eco*RI sequence AATT present at one end of the adaptor. The 5' phosphate residue is generally not present to prevent ligation of the *Eco*RI ends of the adaptors to one another, while the presence of the 5' phosphate on the blunt end allows ligation of the adaptor to another blunt-end DNA fragment. Ligation of the adaptor results in the addition of an *Sma*I site and a cohesive *Eco*RI AATT terminus to the ends of the DNA fragment without requiring digestion of the adapted fragment with *Eco*RI. Since no restriction digestion of the adapted fragment is necessary to free the cohesive ends of the adaptors, this procedure eliminates the need to methylate and protect internal restriction sites.

recombinant DNA methods and the ability to isolate, manipulate, and analyze genes. Although "gene-shaping" has not yet progressed to quite the sophistication imagined by Frank Herbert in his 1966 science fiction novel, development of the ability to modify DNA in a test tube has provided the foundation for the isolation and manipulation of genes.

Summary

1. **DNA fragments can be joined together and specifically modified to produce new DNA molecules.**

2. **DNA ligase is a nick-sealing enzyme that allows the joining of DNA fragments.**

3. **T4 DNA ligase is capable of ligating either cohesive or blunt end fragments and is most frequently used for manipulation of DNA fragments in the test tube.**

4. **The use of other DNA modifying enzymes like Klenow fragment and S1 nuclease helps manipulation of the ends of DNA fragments.**

5. **Short, commercially available DNA fragments called oligonucleotide linkers or adaptors allow specific restriction enzyme sites to be conveniently added to the ends of DNA fragments.**

Vectors

"I get by with a little help from my friends." -- John Lennon/ Paul McCartney

Restriction enzymes give researchers the ability to take a DNA sample containing fragments of random sizes and generate specific fragments. Since these fragments can be separated from one another by gel electrophoresis, it might seem that the restriction enzymes alone are sufficient to allow the isolation of genes. In the simplest strategy, the DNA would be purified, digested with a restriction enzyme, the fragments separated by electrophoresis on a gel, and the desired fragment cut out of the gel.

This scheme will actually work with small genomes like those of plasmids and viruses. However, as the size of the genome increases, several problems complicate this approach. First, as the genome increases in size, more DNA fragments are generated by the restriction enzyme digest. A 5,000 base plasmid genome might give 2 DNA fragments, a 50,000 base bacteriophage genome might yield 25 DNA fragments, while a 5,000,000,000 base eukaryotic genome might generate thousands of individual DNA fragments. As the number of individual DNA fragments increases, the probability that two different DNA fragments will have the same length also increases. Thus, a DNA band on an agarose gel may contain more than one DNA fragment of the same size. In addition, as the number of bands increases, the individual fragments become increasingly difficult to resolve by gel electrophoresis. It is difficult to separate two fragments that are only a few base pairs different in size, causing fragments of very similar size to migrate at the same position in an agarose gel.

A second complication involves the yield of the desired fragment relative to the total amount of DNA. As the size of the genome increases, the relative proportion of the DNA that is the desired gene decreases. A 3000-base gene represents 3000/50,000, or 6%, of a 50,000-base bacteriophage genome. For each 1 milligram of starting DNA, a maximum of 0.06 milligram will be the gene of interest. Assuming no problems with the gel isolation and 100% recovery of the desired DNA, this might be a workable yield of the desired DNA fragment. However, a 3000-base gene represents only 3000/5,000,000,000, or .06%, of a typical eukaryotic genome. A yield of .0006 milligram of desired DNA fragment per milligram of starting DNA is too low to allow isolation and analysis of a specific gene.

A third complication regards the ability to identify the desired DNA fragment in the background of all the other DNA fragments. In the absence of other information, it is not usually possible to simply look at a restriction enzyme digest of a DNA sample and be able tell which DNA fragment contains a desired gene. To be certain that the desired gene has been purified, it might be necessary to purify **each** of the individual DNA fragments in the digested DNA sample. While possible for small genomes, this is a technically and financially unacceptable approach for general use.

Vectors: carrier molecules for other DNA fragments

Although restriction enzymes and DNA ligase can be used to specifically cleave and recombine chromosomal DNA fragments and generate new, recombinant DNA molecules, these capabilities alone are insufficient to allow isolation and characterization of genes. The ability to isolate the desired fragment away from other fragments and to amplify or increase the amount of the pure fragment are fundamental requirements of gene analysis. Both of these needs are provided by specialized helper DNA molecules termed vectors. These vector DNA molecules must be able to provide several specific biological functions to allow convenient propagation of the recombinant DNA molecules. Two properties in particular are crucial to the function of these helper DNA molecules. Vectors must be able to replicate in a host organism to produce many progeny copies of the recombinant molecule and must confer some new phenotypic property so that cells that contain the vector can be distinguished from cells that do not.

Vectors are generally derived from naturally occuring extrachromosomal elements that are capable of replicating in a host (Table 6.1). Many vectors that function in bacteria have been derived from plasmids, small circular DNA molecules that encode various specialized genetic functions, or bacteriophage, such as lambda. The ability to replicate and produce many copies is an important function of the vector, for any DNA fragment that is inserted into the vector will also be replicated, making large quantities of the fragment relatively easy to obtain.

Table 6.1. Examples of bacterial cloning vectors.

Vector	Host	Origin of replication	Selection	Features
pUC19	*E. coli*	ColE1 type plasmid	Ampicillin	General cloning
gt10	*E. coli*	Bacteriophage lambda	Plaques	Insertional
EMBL4	*E. coli*	Bacteriophage lambda	Plaques	Replacement
M13mp19	*E. coli*	Bacteriophage M13	Plaques	Single stranded
pRK290	Gram^{-ve}	RK2 plasmid	Tetracycline	Broad host range
pHV15	*B. subtilis*	pC194 (*S. aureus*)	Cloramphenicol	Shuttle
	or *E. coli*	pBR322 (*E. coli*)	Ampicillin	

Selection of vectors for various host cells

In addition to ability to replicate in a host cell, a vector must also contain a genetic marker that can be used to detect cells that contain the vector (Table 6.1). The most common selective markers present on plasmid vectors (Figure 6.1) are genes that encode resistance to antibiotics, such as

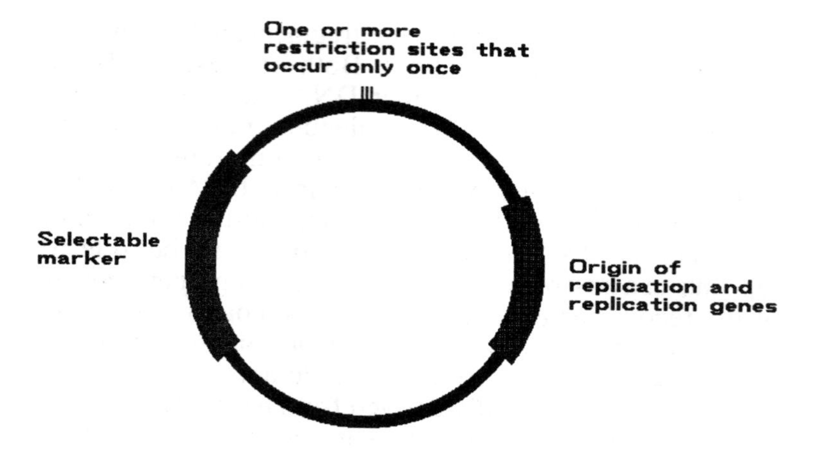

Figure 6.1. A simple cloning vector needs an origin of replication and any genes necessary to allow replication or propagation in a host cell, a genetic marker that confers a property that can be used to identify cells that contain the vector, and one or more restriction endonuclease cleavage sites. Many different types of origin have been used in vectors, including plasmid, chromosomal, bacteriophage, and viral DNA sequences. Many vectors contain genes that confer resistance to antibiotics (ampicillin, chloramphenicol, tetracycline, kanamycin), allowing detection of cells containing the vector while at the same time inhibiting the growth of cells that do not contain the vector. It is important that the restriction sites to be used for DNA fragment insertion be present only once in the vector so that restriction cleavage does not break the vector into more than one DNA fragment. These cleavage sites should also not interfere with the replication or selectable marker regions of the vector.

ampicillin. These markers allow identification of cells that contain the vector and are resistant to the antibiotic. Vectors derived from phage, such as the bacteriophage lambda Charon vectors, do not require an antibiotic resistance gene for genetic detection, for these vectors are capable of forming zones of lysed cells, called plaques, when grown on an indicator lawn of sensitive bacteria.

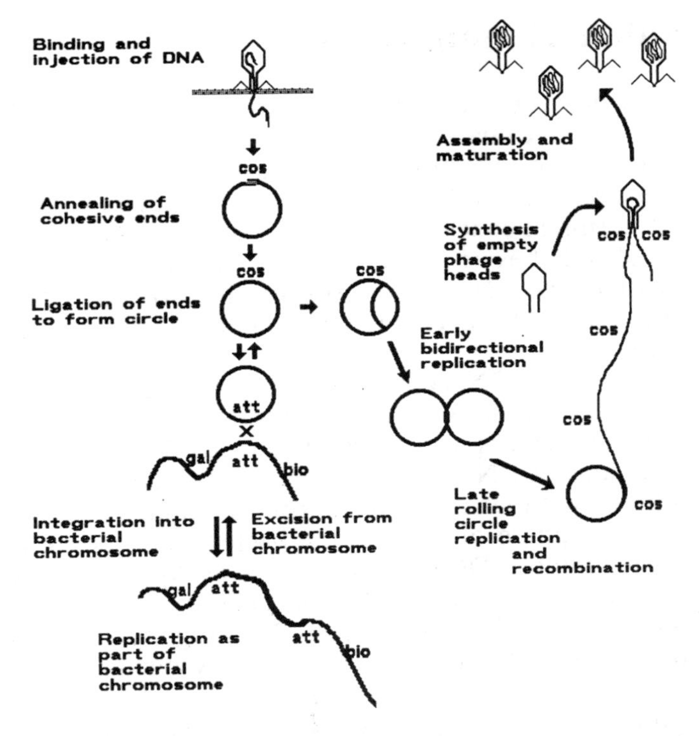

Figure 6.2. Life cycle of bacteriophage lambda. After a bacteriophage particle binds and injects the linear DNA into a host cell, the cohesive ends (*cos*) of the DNA anneal and are ligated to form a circular DNA. This DNA can then lysogenize, or integrate into the bacterial chromosome. A bacteriophage attachment site (*att*) aligns with the bacterial DNA at a site (*att*) located between the *gal* and *bio* genes of *E. coli*. Recombination integrates the circular DNA and the lysogenized lambda DNA is replicated along with the bacterial chromosome. The integrated bacteriophage DNA can be released or excised from the bacterial chromosome to reform the circular phage DNA. Rather than undergoing lysogeny, the circular DNA can commit to a lytic life cycle. The DNA first undergoes bidirectional replication, then converts to a rolling circle mode of replication to generate long end-to-end polymers of newly synthesized phage DNA. Expression of other phage genes results in the synthesis of empty phage heads. The amount of DNA between two *cos* sequences is taken up by an empty head and, if the correct amount of DNA is present in the head (75 to 105% of the lambda genome size), the cos sequences are cleaved to generate the left and right cohesive ends of the linear phage DNA and the assembly process is completed. When sufficient phage assembly and maturation has occurred, lysis of the infected cell releases the progeny phage into the medium.

Lambda cloning vectors

Vectors derived from the bacteriophage lambda take advantage of the dual nature of the normal lambda life cycle (Figure 6.2). The infectious lambda particle contains a linear double-stranded DNA of about 48,500 base pairs with 12-base 5' terminal cohesive (*cos*) ends. After entering a host bacterial cell, the *cos* ends of the lambda DNA anneal with one another and DNA ligase seals the nicks to make a large circular phage DNA. This DNA can undergo two types of life cycle.

The lambda DNA can lysogenize, or integrate directly into the host chromosome. During this lysogenized state, most of the bacteriophage genes are turned off, or repressed, and the lambda DNA functions as a sort of chromosomal hitchhiker without affecting normal cellular processes. During this phase of growth, the bacteria containing the integrated lambda DNA are referred to as "lysogens", and replication of the lambda DNA occurs only as part of the chromosomal replication process. No progeny phage are produced while the phage DNA is in the lysogenized state.

Lambda DNA can also undergo a lytic life cycle in which the circular phage DNA does not integrate into the bacterial chromosome, but enters a two-stage replication process. This replication process results in the synthesis of a a new DNA molecule consisting of many copies of lambda DNA joined end-to-end in a linear multimer. While the phage DNA replicates, expression of genes that code for phage coat proteins results in the assembly of empty bacteriophage heads. As the newly replicated lambda multimer is taken up and packaged into the empty heads, the *cos* sequences on the multimer serve as signals for cleavage of the multimer and final closure of the fully assembled bacteriophage particle.

The genes required for lysogeny of lambda into the host chromosome are conveniently clustered in the central one-third of the lambda DNA molecule, while all of the genes necessary for the lytic life cycle are present on the remaining two-thirds of the lambda genome (Figure 6.3). Restriction enzymes have been used to delete or remove this central "non-essential" region of lambda to construct

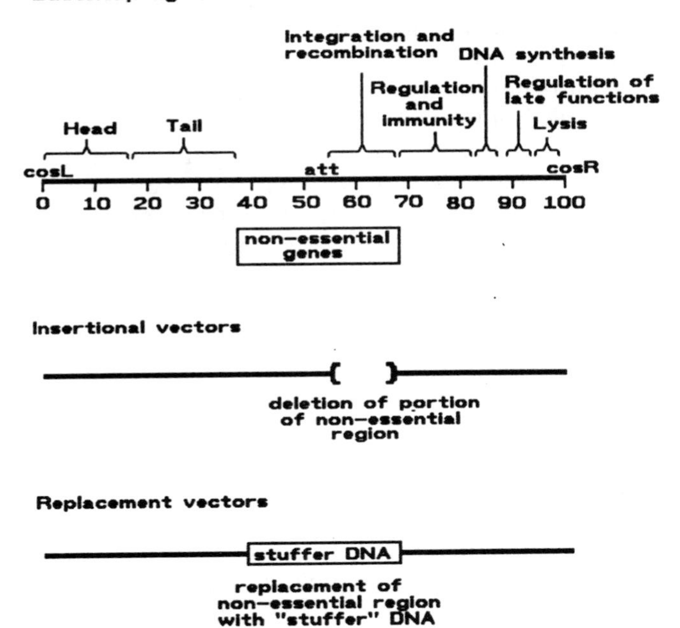

Figure 6.3. Cloning vectors derived from bacteriophage lambda. Relative positions of the genes involved in the major aspects of the life cycle of lambda are shown along the linear DNA map (approx. 48,500 base pairs). The left 40% of the DNA contains the left cohesive end (*cosL*) and genes involved in synthesis of head and tail proteins, while the right 30% contains the right cohesive end (*cosR*) and genes involved in phage replication, in assembly, and in cellular lysis. The central 30% of the genome contains genes involved in lysogeny and are not essential for phage propagation. The empty lambda head can contain a DNA molecule that is at least 75% but not more than 105% of the size of the lambda genome. Insertional vectors contain a deletion that removes a portion of the non-essential region and allows replacement with an additional DNA fragment. Fragments cloned in insertional vectors generally cannot be greater than 9,000 base pairs in size and the vector can be packaged without an extra DNA insert. Replacement vectors have had the entire non-essential region deleted to allow the insertion of larger DNA fragments, generally up to about 23,000 base pairs. Deletion of central region makes the remaining DNA too small to be efficiently packaged, so an extra DNA fragment called "stuffer" is present to allow packaging of the vector. During the cloning procedure, the "stuffer" DNA is removed and replaced with the DNA fragments of interest. Because only recombinants containing extra DNA inserts will be large enough to allow efficient packaging, replacement vectors facilitate the cloning of large DNA fragments.

vectors that allow the insertion of additional DNA fragments. These additional DNA fragments are then propagated during the lytic growth of the vector.

Vectors derived from bacteriophage lambda utilize the fact that the empty lambda head requires a specific amount of DNA - a "headful" - for packaging to occur. DNA molecules where the *cos* sequences are either too close together or too far apart cannot be packaged into empty lambda heads. Two types of vector - insertion and replacement vectors - have been designed to take advantage of this property of lambda DNA packaging.

Insertion vectors have had a portion of the non-essential region cut away with restriction enzymes and replaced with one or more restriction cleavage sites that occur only once in the entire lambda DNA. The purified vector DNA can be digested a restriction enzyme, an additional DNA fragment ligated into the vector, and the ligation products packaged using a preparation of empty lambda phage heads called a "packaging mix". Because there is a maximum length of DNA that can be picked up by the empty heads, there is a limit to the size of DNA fragment that can be inserted in a lambda insertion vector, and the maximum allowable insert size is generally about 9,000 base pairs.

Replacement vectors have been constructed by using restriction enzymes to remove the entire non-essential region of lambda. Since removing this much DNA makes the lambda derivative too small to package in empty phage heads, an extra DNA fragment called "stuffer DNA" has been substituted in place of the non-essential DNA to allow efficient packaging of the vector DNA. To construct a recombinant DNA molecule, the vector DNA must be digested with a restriction enzyme that releases the stuffer DNA fragments and the vector "arms" purified away from the stuffer DNA. The purified vector "arms" can then be ligated with the desired additional DNA fragment and packaged into empty phage lambda heads. Replacement vectors take advantage of the fact that the vector arms ligated together without an insert are too small to be packaged in empty phage heads, while lambda arms ligated with an extra DNA fragment can be large enough to allow efficient packaging. These replacement vectors preferentially select for recombinants containing certain size classes of DNA fragments and can accept total inserts as large as 20,000 base pairs, a size that is difficult with other vectors.

Vectors derived from the single-stranded bacteriophage M13

A series of vectors has been derived from the single-stranded DNA bacteriophage M13 to take advantage of the unusual life cycle of this bacterial virus (Figure 6.4). M13 is a male-specific bacteriophage that only infects bacteria that produce an F-pilus, the mating apparatus used in bacterial conjugation and transfer of the F-factor, a large, low-copy, plasmid that can be sexually transmitted to other bacteria. The infectious M13 phage particle contains a single-stranded, circular DNA molecule that forms a double-stranded circle upon entry into a host bacterium. This double-stranded DNA circle becomes the replicating form (RF) and produces more double-stranded copies. The double-stranded circles also serve as a template for the production of single-stranded circular DNA molecules that are the same as the single-stranded circular DNA that originally infected the host cell.

This single-stranded circular phage DNA is packaged in bacteriophage coat protein and extruded from the cell into the medium as an infectious bacteriophage particle without causing the infected cells to lyse. Vectors derived from M13 are particularly useful because the infected cells serve as a source of the double-stranded RF DNA and the extruded bacteriophage particles serve as a source of the single-stranded circular DNA. The double-stranded circular RF DNA can be purified from the infected cells and treated with restriction enzymes and DNA ligase just like a plasmid vector DNA to allow the construction and analysis of recombinant molecules. In addition, the single-stranded circular bacteriophage DNA can be purified from the infectious particles and used for specific experimental procedures that require single-stranded DNA, such as nucleotide sequence analysis.

Insertional inactivation for detection of DNA inserts

Many vectors also contain a second genetic marker that uses the property of insertional inactivation to determine whether the DNA molecule contains an extra DNA fragment (Figure 6.5). This second marker gene has generally been engineered to contain a multiple cloning site (MCS), a series of restriction endonuclease sites that each occurs only once

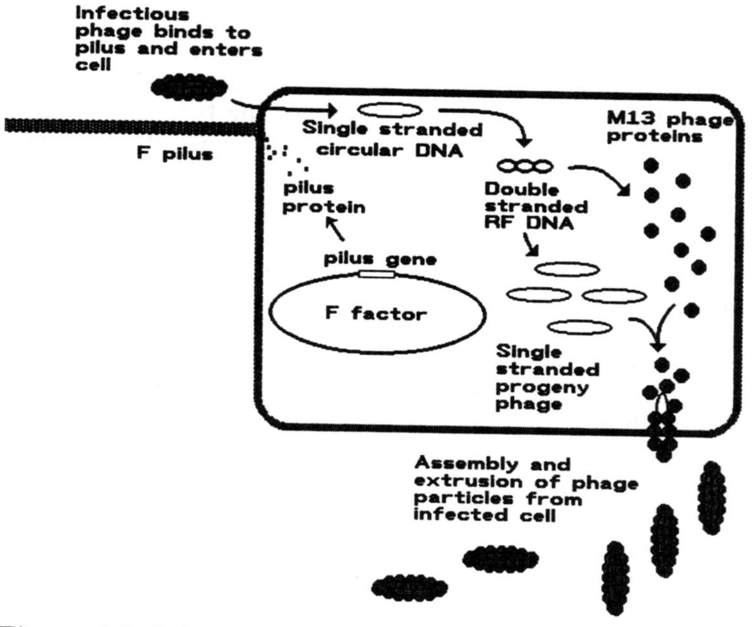

Figure 6.4. Life cycle of the male-specific, single-stranded bacteriophage M13. M13 requires the presence of the bacterial F pilus for infection of a host cell. The pilus gene is normally present on the F factor, a large self-transferable plasmid. Upon entry of the bacteriophage particle, the single-stranded circular phage DNA is converted to a double-stranded replicating form (RF). The double-stranded RF both replicates and produces single-stranded progeny phage DNA. Expression of phage genes also results in the synthesis of M13 coat proteins. The newly synthesized single-stranded progeny phage and the coat proteins assemble and are extruded through the infected cell membrane into the medium. Unlike cells infected with bacteriophage lambda, cells infected with M13 do not lyse but remain viable and continue to grow, although at a somewhat slower rate than uninfected cells.

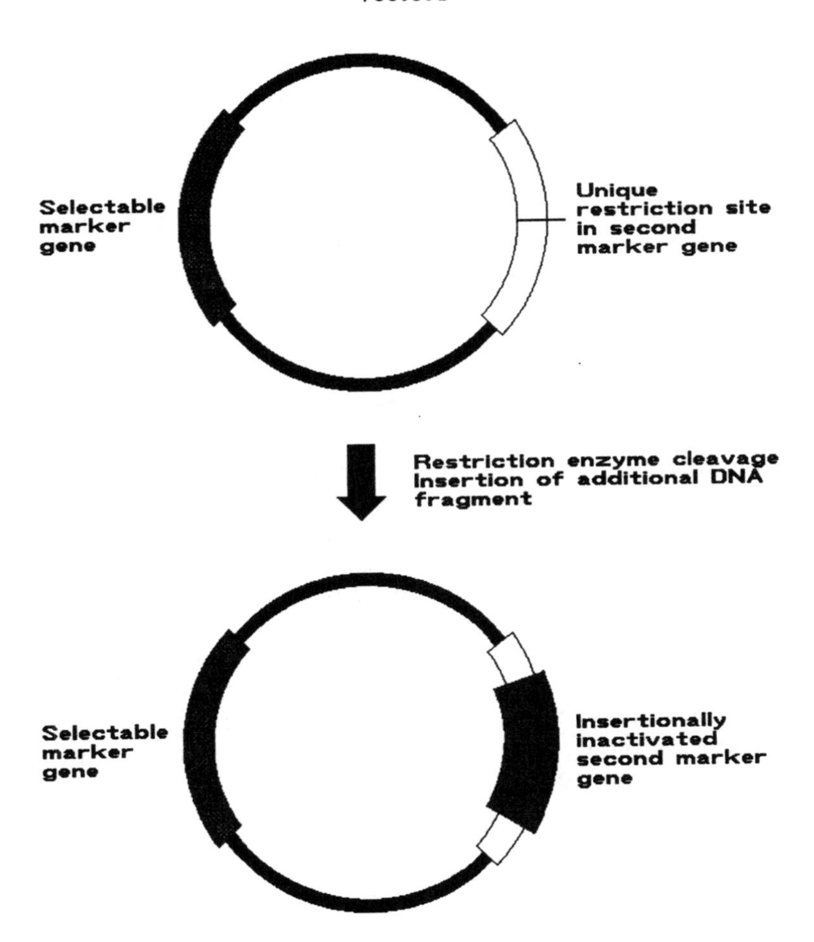

Figure 6.5. Insertional inactivation as a method for detecting DNA inserts. DNA inserts in the cloning vector are detected by a change in genetic properties that accompanies the insertion of the DNA fragment. The vector must contain a selectable marker gene, such as ability to confer resistance to an antibiotic like ampicillin or ability to form a plaque on a suseptible host, that remains unaffected by the cloning process. A second marker gene must contain a restriction cleavage site that occurs only once in the vector. Insertion of an additional DNA fragment in this cleavage site will interrupt the second marker gene and prevent the function of the second marker gene. The wild type vector will give the phenotype [Marker 1^+ Marker 2^+] and the recombinant vectors will give the phenotype [Marker 1^+ Marker 2^-]. This allows the recombinants to be easily distinguished from the original vector.

in the vector DNA. The MCS allows the vector DNA to be cleaved to a linear molecule with a restriction enzyme and facilitates the efficient insertion of additional DNA fragments. A DNA fragment that inserted into one of the MCS restriction sites will interrupt the marker gene and prevent the expression of the phenotype normally associated with that gene. A recombinant vector can be distinguished from the original vector by the loss of function of the marker gene.

Plasmid vector pUC19

The plasmid vector pUC19 (Figure 6.6) contains a replication origin derived from a high-copy plasmid to allow replication of the vector in *E. coli* and a gene that produces ß-lactamase, an enzyme that inactivates ampicillin, and confers resistance to ampicillin. This vector also contains a portion of the *E. coli* ß-galactosidase gene that produces a protein known as the ß-galactosidase α-donor. The remaining receptor portion of the ß-galactosidase protein can be produced by a modified ß-galactosidase gene that is present on the *E. coli* chromosome. Although neither the α-donor nor the receptor protein is enzymatically active by itself, the proteins can combine with each other to form an active ß-galactosidase complex. When cells containing the vector pUC19 are exposed to the chemical inducer isopropylthio-ß-galactoside (IPTG), the ß-galactosidase genes are expressed and the α-donor and receptor proteins are synthesized. These proteins associate with each other to form an active ß-galactosidase protein that can convert the chromophore 5-bromo-4-chloro-3-indolyl-ß-D-galactoside (X-gal) to a blue compound. Thus, in the presence of IPTG and X-gal, cells containing the vector pUC19 form ampicillin-resistant, blue colonies.

The series of unique restriction endonuclease cleavage sites that forms the MCS in the pUC19 is located in the ß-galactosidase α–donor gene. Insertion of DNA fragments in the MCS interrupts the gene and prevents the production of the ß-galactosidase α–donor protein. Recombinant plasmids containing DNA inserts remain capable of conferring resistance to ampicillin, but are no longer able to produce the blue color in the presence of X-gal. This property makes it easy to distinguish cells that contain pUC19 and form ampicillin-resistant blue colonies from cells that contain recombinant pUC19 derivatives and form ampicillin-resistant white colonies.

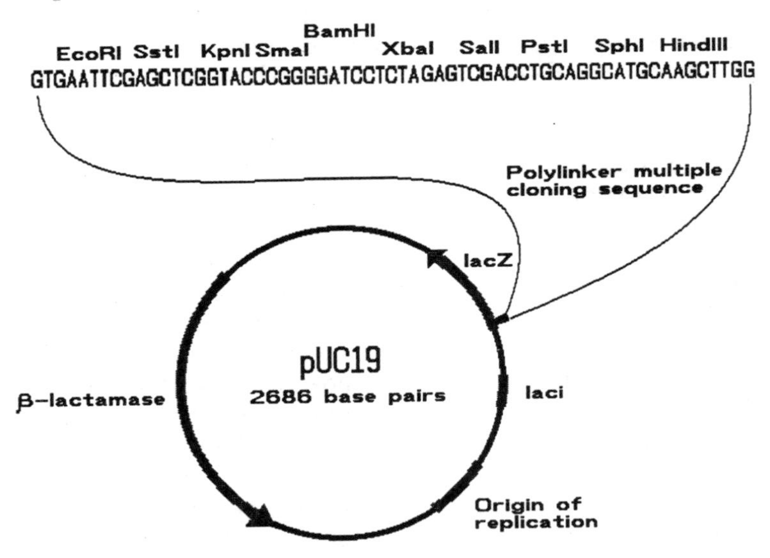

Figure 6.6. Restriction map of the plasmid cloning vector pUC19. The important features of this plasmid include an origin of replication that allows high-copy replication in *E. coli* and closely related bacteria, a gene encoding the enzyme ß-lactamase, which confers resistance to ampicillin, the *laci* regulatory region of the *lac* operon of *E. coli* and a modified *lac*Z gene containing a polylinker multiple cloning sequence. The *lac*Z gene produces a derivative of the protein ß-galactosidase that can, in the correct bacterial hosts, cause the production of a blue compound from the galactose derivative called X-gal. This plasmid can cause the formation of blue colonies when present in cells grown on medium containing both ampicillin and X-gal. The beginning of the *lac*Z gene (the aminoterminal end) contains a polylinker multiple cloning sequence, shown above the plasmid. Each of the restriction enzyme sites shown in this sequence occurs only once in the plasmid. Insertion of an extra DNA fragment into one of these cleavage sites will interrupt the *lac*Z coding region and prevent the synthesis of the ß-galactosidase protein. While recombinant plasmids will retain the ability to confer resistance to ampicillin, cells containing recombinants will be unable to convert X-gal to the blue compound and colonies will remain white.

The modified ß-galactosidase gene is probably the most common marker gene used for detection of DNA inserts by insertional inactivation. This gene derivative has been used in a wide variety of vectors derived from both plasmids and bacteriophage. Vectors derived from M13, such as M13mp19, for example, use the same detection principle as pUC19, with minor modifications to suit detection of the bacteriophage. Because M13 derivatives require the presence of the F-pilus for the phage to be able to infect the host cells and form plaques, the bacterial hosts used to propagate M13 derivatives usually contain an F-factor. A genetic trick has been used to ensure that the F plasmid remains in the bacteria during growth. A deletion that inactivates the genes that synthesize the amino acid proline was introduced into the bacterial chromosome to make the bacteria *pro*$^-$ and unable to grow in the absence of proline. An F plasmid that contained the genes that allow synthesis of proline was then introduced into the *pro*$^-$ host to make the bacteria able to grow in the absence of added proline. As long as the bacteria are maintained on mimimal medium containing no extra proline, only cells that contain the F plasmid will be able to grow. Since all the growing cells contain the F plasmid, they must also all produce the F-pilus and can therefore be infectable by M13 derivatives.

Insertional inactivation in M13 vectors

Vectors derived from M13, like M13mp19 (Figure 6.7), generally contain the same ß-galactosidase donor gene that is present in the plasmid pUC19, and the bacterial host contains the ß-galactosidase receptor gene. When M13mp19 infects an appropriate host, such as *E. coli* JM105, and IPTG and X-gal are added to the medium used to grow the bacteria, the plaques that result from infection of the bacteria with the bacteriophage DNA will turn blue. As is the case with the plasmid pUC19, the M13mp19 MCS is in the ß-galactosidase α–donor gene. DNA fragments inserted in the M13mp19 MCS also disrupt the production of the α–donor protein and plaques resulting from recombinant M13 derivatives are unable to turn blue. Thus, plaques that are

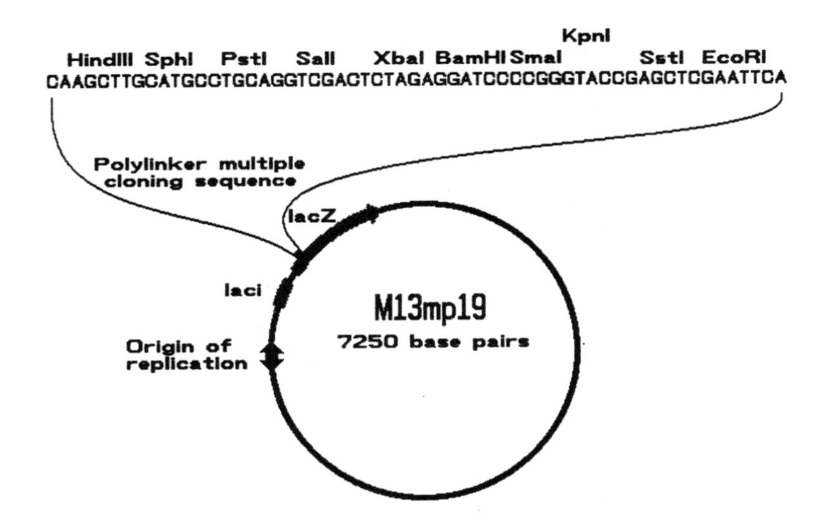

Figure 6.7. Restriction map of the cloning vector M13mp19. This vector contains the same *laci*/*lac*Z region as the plasmid vector pUC19 but contains the origin of replication and all of the genes(not shown) of bacteriophage M13. M13mp19 will cause the formation of blue plaques, or zones of infected cells, on a lawn of bacteria grown in the presence of X-gal. Insertions of DNA in the multiple cloning sequence interfere with ability to produce ß-galactosidase and cause the formation of clear or white plaques. As is the case for the plasmid pUC19, M13mp19 recombinants can be distinguished from the original vector by inability to produce a blue color from X-gal.

caused by recombinant M13 derivatives remain clear or white and can be readily distinguished from the blue plaques caused by the M13 vector alone.

Host range of vectors

When introducing DNA into bacteria other than *E. coli*, it is important to realize that many of the cloning vectors specifically designed for use in *E. coli* contain genes that are not functional in other bacteria. The origin of replication that is present on the plasmid vector pUC19, for example, only allows replication of the vector in bacteria that are very closely related to *E. coli* and the vector cannot replicate in most other Gram-negative or Gram-positive bacteria.

Broad host range vectors are derived from plasmids like RK2, RSF1010, and pSa that are naturally capable of replicating in a wide variety of Gram-negative bacterial hosts. Although these plasmids can replicate in many different bacteria, they are not capable of replicating in Gram-positive bacteria like *Bacillus subtilis*. Composite or shuttle vectors that contain two different origins of replication, one that allows replication of the vector in *E. coli*, and a second that allows replication in a second host, have been developed to allow DNA fragments to be cloned in *E. coli*, then efficiently introduced back into and maintained in another host.

Vectors have also been designed that allow cloning of DNA fragments in eukaryotes. Cloning in yeast is usually performed with special shuttle vectors that contain a plasmid origin of replication to allow growth and manipulation of the vector in *E. coli* and an origin to allow replication in yeast. Three types of yeast replication origin have been used in the construction of yeast cloning vectors. YEp vectors have a yeast origin derived from the naturally occurring yeast 2 micron plasmid and are maintained at high copy (50-100 copies/cell) in yeast. YRp plasmids, which contain an autonomously replicating sequence (*ars*) isolated from yeast, can be transformed very efficiently into yeast, but are relatively instable and are easily lost from the cells. The YCp plasmids contain a yeast *ars* sequence and part of a yeast centromere, the portion of the chromosome involved in efficient separation of daughter chromosomes during cell division, and are much more stable than YRp plasmids.

As has been done with prokaryotes and simple eukaryotes, vectors have been constructed that allow DNA to be stably maintained extrachromosomally in higher eukaryotic cells. These vectors are generally derived from viruses that normally infect and propagate in the desired cell type, but infectious aspects of the virus have usually been removed and replaced with a selectable marker like the neomycin phosphotransferase gene or the herpes simplex virus thymidine kinase gene. The mammalian viruses SV40, polyoma, bovine papillomavirus, and vaccinia, for example, have all been used for the

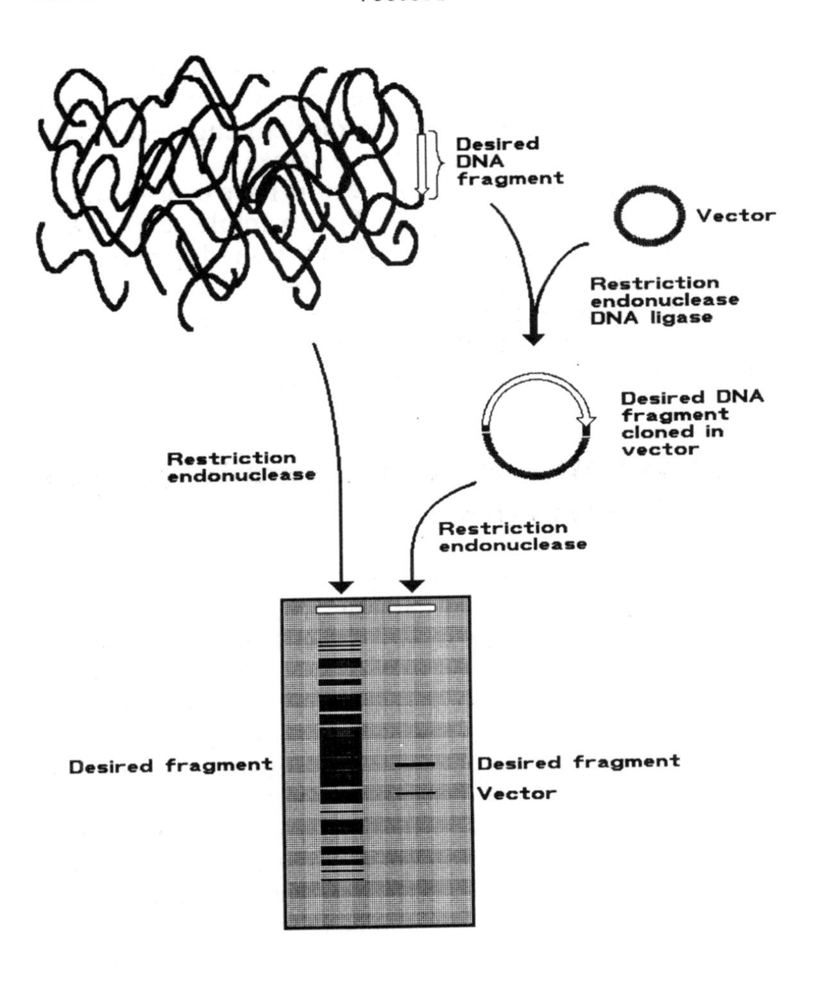

Figure 6.8. Cloning a DNA fragment in a vector as a means of isolating a desired fragment. A restriction endonuclease can be used to digest a chromosomal DNA sample to generate a specific fragment that contains a desired gene. This fragment may be one of several hundred individual fragments and electrophoresis may not sufficiently resolve the desired fragment to allow purification away from other fragments of similar size. When the desired fragment is inserted into a cloning vector, restriction endonuclease digestion and gel electrophoresis will release the desired insert from the vector. The desired DNA fragment is effectively purified away from the undesired fragments during the cloning procedure.

construction of vectors that will propagate in mammalian cells. While the use of these vectors is somewhat more complex than the use of prokaryotic cloning vectors, the principles remain substantially the same.

Many types of "specialty" vectors have been designed to accomplish specific goals. For example, expression vectors allow the over-production of a desired gene product from a cloned DNA fragment. All these specialty vectors utilize the same principles discussed for other vectors, but have been additionally designed to allow efficient manipulation or analysis of DNA fragments.

Vectors both isolate and amplify the yield of a DNA fragment

Vectors have proven invaluable in the isolation and characterization of DNA fragments. When a desired DNA fragment is ligated into a vector, the DNA fragment has been isolated away from other DNA fragments in the original DNA preparation. The recombinant DNA molecule can be purified and digested with the same restriction enzyme used to construct the recombinant, releasing the inserted DNA fragment from the vector. If the desired DNA insert is the same size as the vector alone, the relative yield of the desired fragment will be 50%, a significant increase over the .06 to 6% yield obtained when the fragment is purified directly from chromosomal DNA. The recombinant vector containing the extra DNA insert can be used as a source of large amounts of the purified desired fragment.

The insertion of a DNA fragment into a vector (Figure 6.8) accomplishes two basic gene isolation requirements: the desired DNA fragment is now present on a replicating DNA molecule that can be propagated in a host to produce large amounts of the fragment, and the fragment has been efficiently isolated away from all of the other undesired DNA fragments to increase the relative yield of the desired fragment. The DNA fragment present in the vector is often called a molecular clone or a cloned DNA fragment and the process of generating the recombinant is usually referred to as "cloning".

Summary

1. **Vectors are DNA molecules that can reproduce themselves in a host cell and can accomodate the insertion of extra DNA.**

2. **Vectors allow the production of large amounts of a desired fragment.**

3. **Insertional inactivation of a marker gene allows a recombinant vector to be differentiated from a non-recombinant vector.**

4. **Vectors allow the molecular cloning of DNA fragments away from other fragments in a preparation.**

Transformation

"With an anxiety that almost amounted to agony, I collected the instruments of life around me, that I might infuse a spark of being into the lifeless thing that lay at my feet. It was already one in the morning; the rain pattered dismally against the panes, and my candle was nearly burnt out, when, by the glimmer of the half-extinguished light, I saw the dull yellow eye of the creature open; it breathed hard, and a convulsive motion agitated its limbs." — Frankenstein or The Modern Prometheus, Mary W. Shelley

Contrary to the perception that the recombinant DNA molecules that result from *in vitro* manipulation of DNA give rise to new life forms, these new DNA molecules are lifeless bits of nucleic acid. They cannot propagate or reproduce themselves any more readily than can a grain of sand. However, because nucleic acids are virtually universally recognized as genetic information by living cells, these new DNA molecules can, if introduced into and maintained in an already living cell, confer new phenotypic traits on the host cell. It would be more appropriate to say that these new molecules **modify** existing organisms rather than generate new life forms.

Because the new recombinant molecule cannot reproduce by itself, once a DNA fragment has been ligated to a vector *in vitro*, it is necessary to introduce the construct into a host where it can replicate and produce many copies of itself. The process of introducing a DNA molecule into a host cell is called transformation; when the introduced

DNA is a type of viral or bacteriophage DNA, the process is often referred to as transfection. The transformation step and the subsequent biological amplification of the inserted DNA molecule was the second major innovation of recombinant DNA methodology that was critical to improving the ease of the isolation and analysis of specific DNA molecules.

Competency and transformation: the ability to take up DNA

Although certain types of bacteria have DNA transport systems that will take up DNA, cells generally must be treated to make them competent for DNA uptake. The most commonly used host for the propagation of recombinant molecules is the bacterium *Escherichia coli*. *E. coli* can be made competent for DNA uptake by treating a rapidly growing culture of the cells with a high concentration of the salt calcium chloride (Figure 7.1). This causes the cells to swell and become sphereoplasts capable of binding DNA. When placed at 42°C for 1-2 minutes, the sphereoplasts contract and the bacteria take up any bound DNA. The cells containing the DNA can then be plated on a growth medium that selects for cells containing the plasmid vector and inhibits the growth of other cells. This selection is usually accomplished by adding an antibiotic like ampicillin to the growth medium. Following a period of incubation to allow growth, colonies of transformed bacteria will form on the growth medium. These transformants must then be purified and the DNA present in the colonies extracted and characterized.

Transfection

Transfection systems involving viral DNA are somewhat different in that the viral DNA may be either inserted into competent cells or packaged in an empty viral head to form an active viral particle capable of infecting a host cell

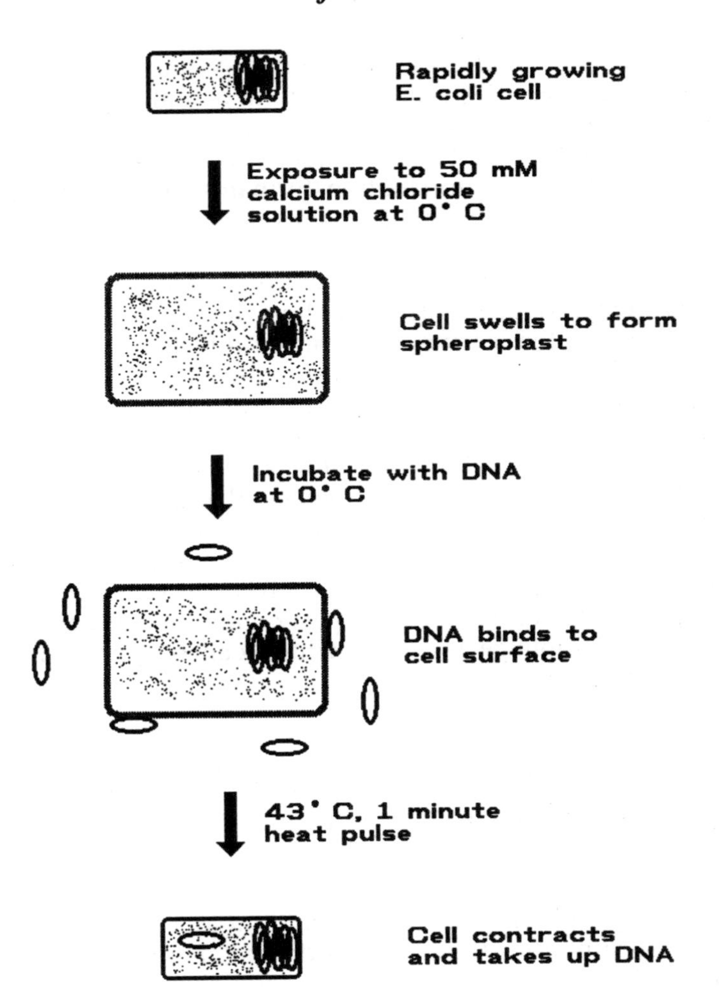

Figure 7.1. Calcium chloride transformation of *E. coli*. Rapidly growing bacteria are harvested by centrifugation and resuspended in a solution of 50 mM calcium chloride. During a short incubation on ice, the cells swell and become spheroplasts. DNA that is then added to the sphereoplasts will stick to the outer surface of the cells. DNA uptake is caused by a brief heat pulse, which causes the cells to contract and take up the bound DNA. A variety of transformation protocols exist that modify the concentration or type of salt (for example, some protocols use rubidium chloride), but the basic transformation principle remains the same. The chemical treatment causes the cells to become able to take up added DNA.

(Figure 7.2). While the packaging systems require the additional step of mixing DNA with a solution of empty bacteriophage lambda heads, the efficiency of introduction of DNA of high molecular weight (>20,000 base pairs) is much higher. Selection for transformants containing a viral vector is also different than selection for transformants containing plasmid vectors. Detection of the cells that take up the bacteriophage DNA generally uses the ability of the viral vector to form a plaque, or hole, in a layer of indicator bacteria, called a lawn, that can be infected by the viral vector. The transformed or transfected bacteria must be mixed with a sample of the indicator bacteria, spread on the surface of a solid growth medium, and incubated to allow the growth of the indicator lawn. Transformed cells will produce progeny bacteriophage that are released into the bacterial lawn, where they can attach to and infect more bacteria. As the lawn grows, clear spots or plaques in the lawn form where bacteriophage have infected the host indicator bacteria. These plaques contain many copies of the bacteriophage progeny, which can be purified and characterized.

The bacterium *Escherichia coli* and routine cloning

Most routine cloning is done in *E. coli*. Although the use of this very well-characterized bacterium as a host organism greatly simplifies the construction and characterization of recombinant DNA molecules, genetic characterization of isolated genes may require that the DNA be re-introduced back into the original organism from which the gene was isolated (Figure 7.3). For example, a recombinant plasmid that contains a host specificity gene isolated from the Gram-negative, root-nodulating soil bacterium *Rhizobium trifolii* may not function properly in an *E. coli* host strain. It may be necessary to transfer the recombinant plasmid from *E. coli* to *R. trifolii* to examine the function of the isolated *R. trifolii* gene. Transformation procedures have been developed that allow the introduction of DNA into a wide

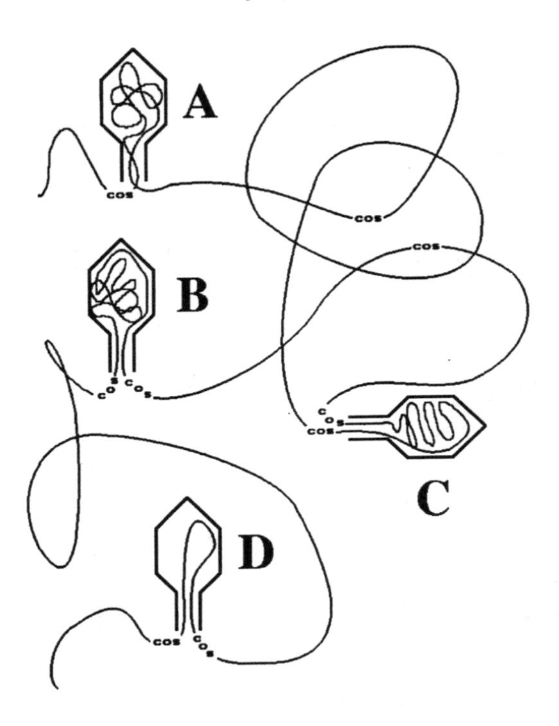

Figure 7.2. Packaging of bacteriophage lambda DNA by empty bacteriophage heads. The long polymer of lambda DNA that is produced during the rolling circle phase of replication, indicated by the *cos* sequences located along the DNA strand in the figure, serves as a template for empty bacteriophage capsids or heads. The empty head must take up sufficient DNA located between two *cos* sequences to have 75 to 105% of the amount of DNA present in the normal lambda bacteriophage genome. If the correct amount of DNA is taken up, the two *cos* sequences will be correctly positioned for cleavage of the DNA to generate the single-stranded cohesive ends of the bacteriophage and the head will mature to form an infectious particle. The heads shown in B and C both have sufficient DNA and aligned *cos* elements and will mature correctly. The phage head shown in A has taken up DNA located between two *cos* elements that are located too far apart and maturation will not proceed. The phage head shown in D has taken up DNA between *cos* elements that are too close together and will also be unable to form an infectious bacteriophage particle. Empty bacteriophage head preparations or packaging mixes can be used to package any DNA that contains *cos* elements located the correct distance apart. During packaging of a lambda vector that has been ligated to other DNA fragments, only the ligation products of the correct size will be taken up by the empty bacteriophage heads to form infectious particles that can inject the packaged DNA into host bacteria.

variety of both Gram-negative and Gram-positive bacteria. The use of a shuttle vector that can replicate in either *E. coli* or another host allows the transformation of DNA purified from *E. coli* into another host for genetic examination.

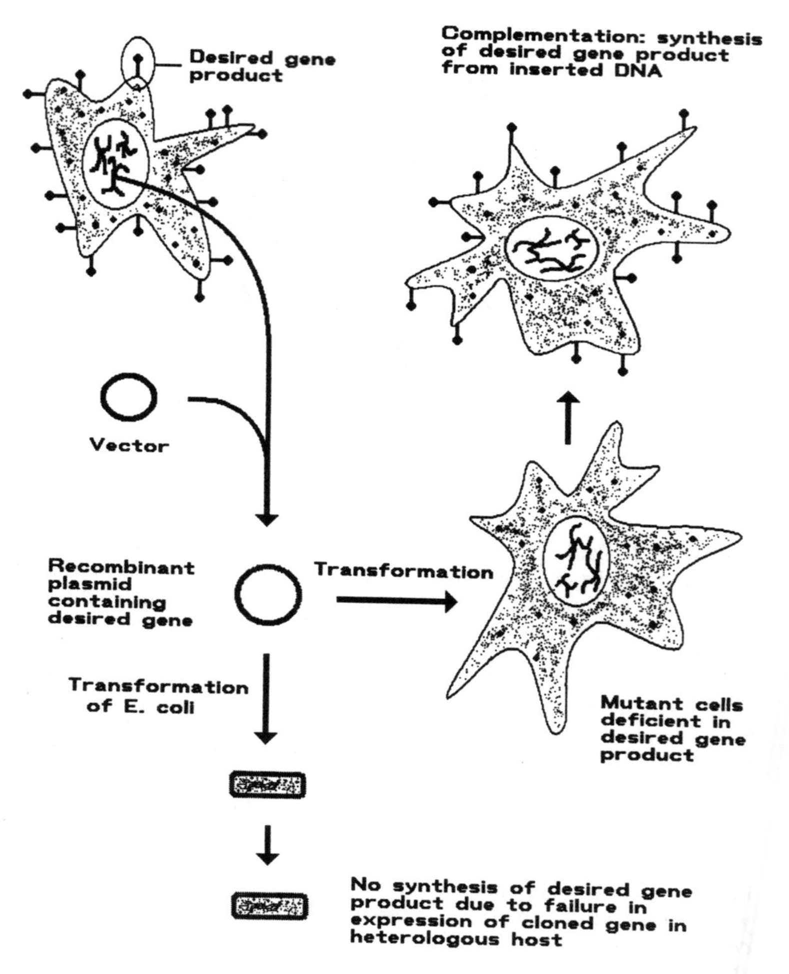

Figure 7.3. Genetic analysis of cloned DNA often requires that the isolated gene be re-introduced into the host from which it was originally isolated. When the cloned gene is transformed into a cell type that is capable of expressing the gene but has been mutated or made deficient in the desired gene product, the gene will be expressed and the desired gene product will be detected in the transformed cells.

Conjugation as an alternate means of introducing DNA

Unfortunately, not all bacteria can be efficiently transformed with DNA. Special vectors have been constructed that use normal bacterial conjugation or sexual mating systems to allow the transfer of DNA from *E. coli* to a second type of bacteria (Figure 7.4). In addition origins of replication that function in *E. coli* and in the desired nontransformable bacterial host, these vectors contain an origin of transfer (*ori*T) that allows the vector to be mobilized and transferred to other bacteria. The genes that promote mobilization and conjugal transfer are often maintained on a second transfer helper plasmid that only replicates in *E. coli*. Transfer of the vector is accomplished in a tripartite, or three-way, conjugation. *E. coli* containing the vector to be mobilized, *E. coli* containing the transfer helper plasmid, and the recipient bacteria are all mixed together, generally on the surface of a solid nutrient medium, and allowed to grow. The transfer helper plasmid transfers itself into the bacteria containing the vector, where the vector *ori*T is recognized by the helper plasmid transfer functions, mobilizing the vector. The vector then transfers to the non-*E. coli* recipient bacteria. The mating mixture can then be placed on medium containing antibiotics to select for the growth of the transformed bacteria. This scheme allows the introduction of DNA into bacteria that cannot be easily transformed by conventional methods.

Transformation of eukaryotes

Transformation of simple eukaryotes, like yeast, can often be accomplished with chemical treatments very similar to those used for transforming bacteria. A variety of other methods, including DNA uptake, electroporation, and lipofection, have been developed for introducing DNA into higher eukaryotes. Mammalian cells grown in a culture dish, for example, can be transformed with DNA that has been allowed to form a precipitate with calcium phosphate.

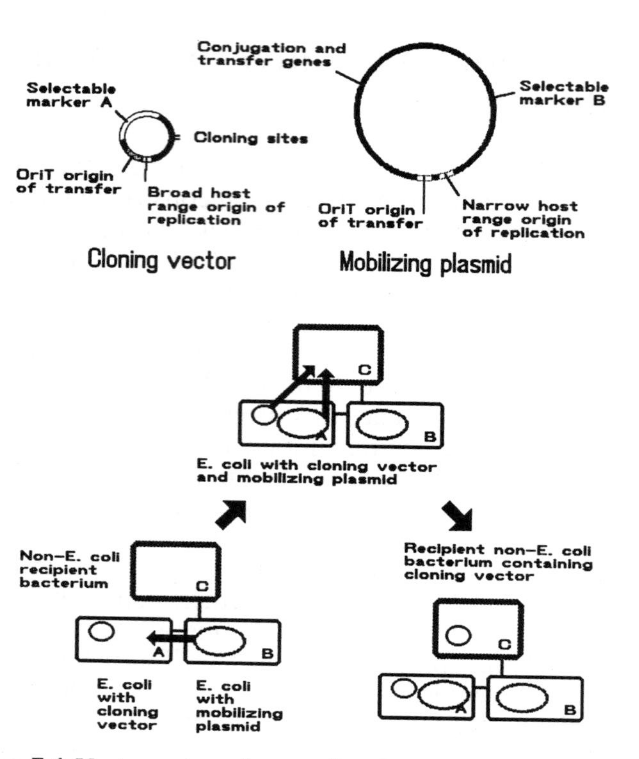

Figure 7.4. Vector systems for transfer of genes to bacteria that cannot be transformed with DNA. A small cloning vector contains and origin of replication that allows the plasmid to replicate in a wide variety of bacterial hosts, a selectable (generally conferring resistance to an antibiotic), one or more unique restriction sites for the insertion of DNA fragments, and an origin of transfer (*ori*T) that allows the plasmid to be mobilized and transferred during conjugation. The conjugation and transfer genes are generally present on a mobilizing plasmid that has an origin of replication allowing maintenance in a narrow range of hosts. To transfer the cloning vector to a bacterium that cannot be readily transformed, the strain of *E. coli* containing the vector (A), the strain of *E. coli* containing the mobilizing plasmid (B), and the recipient bacteria (C) are mixed together. A three-way, or tripartite, mating occurs and the mobilizing plasmid transfers to the bacteria containing the vector. It then promotes transfer of the vector as well as itself, and both plasmids can transfer to the non-*E. coli* recipient bacteria (C). The vector with the broad host range origin can continue to replicate in the new host, but the mobilizing plasmid with the narrow host range origin cannot replicate in the new host and is rapidly lost. Antibiotic selection or nutrient conditions are generally used to eliminate the *E. coli* cells (A and B) and select for the non-*E. coli* bacteria (C) containing the vector.

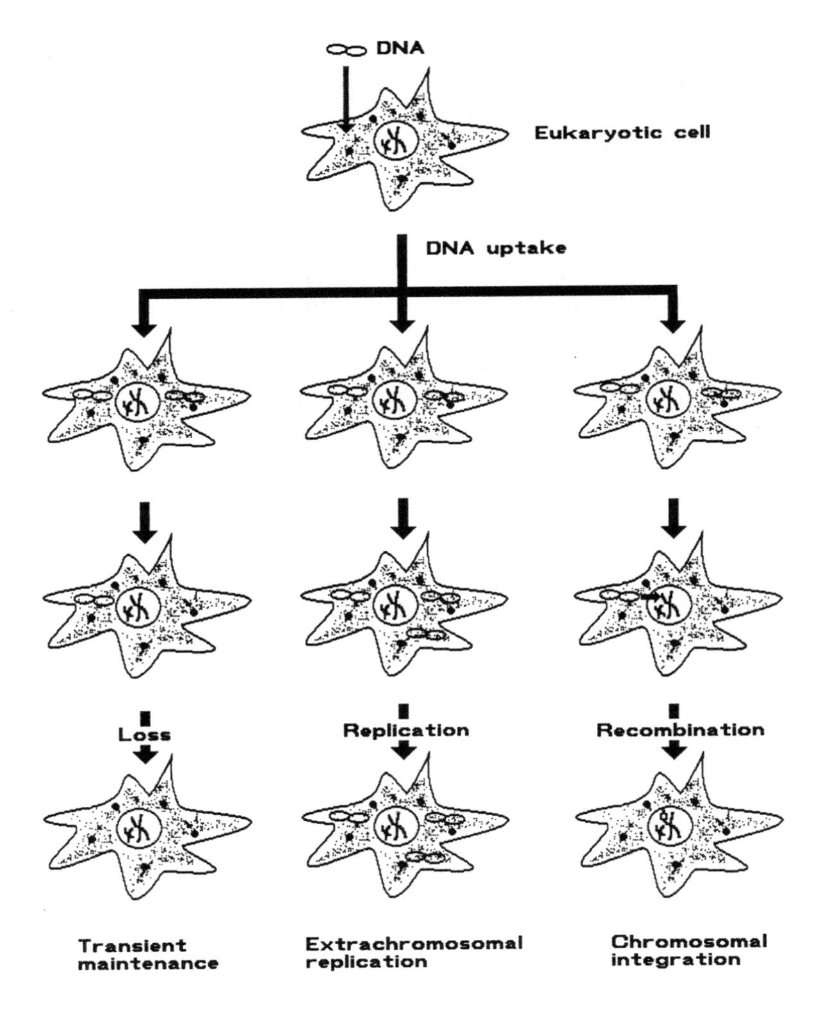

Figure 7.5. Potential fates of DNA transformed into a eukaryotic cell. DNA uptake can be followed by transient maintenance, with the introduced DNA maintained in the cell for a period of hours to several days before the DNA is degraded and lost. During this time, the DNA can be expressed and gene products detected. The addition of an origin of replication to the added DNA allows the DNA to be maintained extrachromosomally. The location in the cell, the number of copies that are present, and the relative stability of the DNA can all be affected by the replication sequences present on the DNA. Added DNA can also undergo recombination that incorporates the DNA into the host chromosome, where the added DNA is subject to normal chromosomal replication and gene expression processes.

Electroporation is a fairly efficient transformation technique that uses a brief exposure to high voltage to cause cells to take up DNA through the cellular plasma membrane. Lipofection makes use of small lipid vesicles or droplets complexed with DNA to fuse with cells and introduce the DNA.

DNA taken up by a eukaryotic cell can undergo one of three basic fates: it can be transiently maintained, maintained extrachromosomally with the use of a cloning vector, or integrated into the genomic DNA (Figure 7.5). DNA that is introduced into a mammalian cell is transiently maintained and can persist for many hours in the absence of replication or integration of the added DNA. This period of time is often sufficient to allow measurment of gene function. Transient expression assays measure the function of the introduced DNA in the short period of time after the DNA is introduced into the cell (typically within 48 hours of transformation) and do not require the DNA to be replicated or stably integrated into the host chromosome. These assays play an important role in the analysis of regulation of eukaryotic gene expression because they allow investigation of the function of introduced DNA in the absence of requiring that the DNA be stably maintained.

DNA that is introduced into a cell can also undergo recombination and be integrated into the host chromosome. Homologous recombination occurring between chromosomal DNA sequences and similar sequences present in the introduced DNA can cause the chromosomal integration of all or a portion of the transformed DNA molecule. A single recombination event, often called a cross-over, between a circular plasmid DNA and the host chromosome incorporates the entire circular DNA. A pair of cross-over events in the same gene will exchange the sequences located between the two cross-over points (Figure 7.6). Once integrated into the chromosomal DNA, the introduced DNA is maintained by replication of the host chromosome.

Homologous recombination has been used to introduce a specifically modified gene back into the host chromosome to test genetic function of the modified gene. A gene can be isolated from the original host and manipulated in *E. coli*, but the effects of alterations or mutations in the gene must be examined in the cell type from which the gene was originally obtained. Homologous recombination can be used to integrate a desired altered gene back into the chromosome where it will be subject to normal cellular regulatory processes.

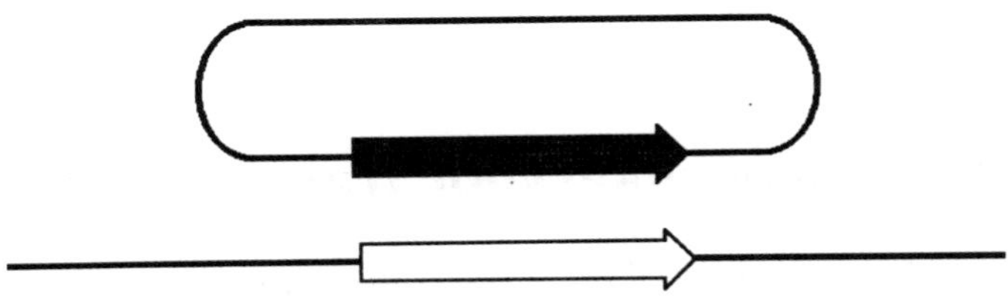

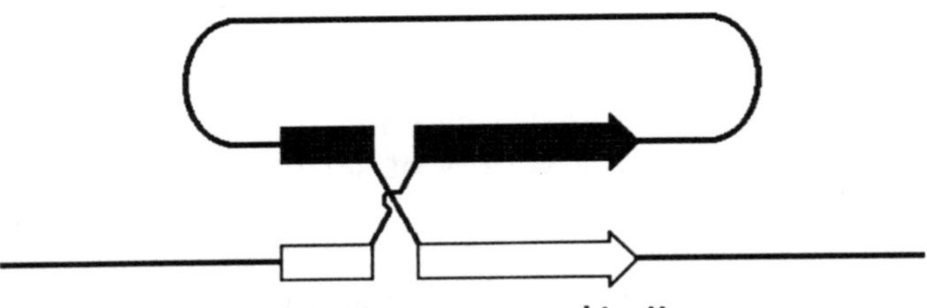

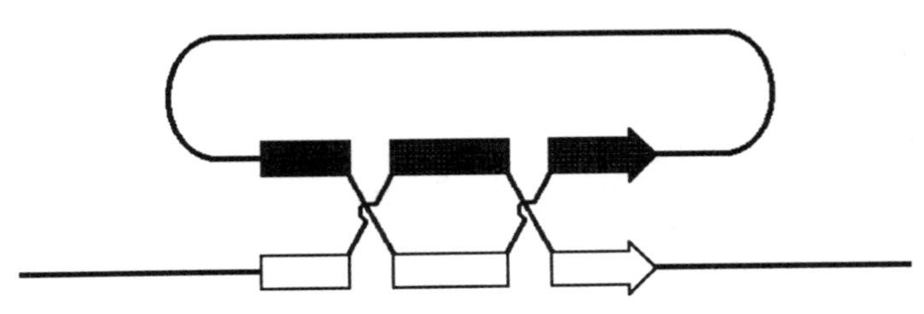

Figure 7.6. Recombination processes can introduce DNA into the host chromosome. A single recombination event can introduce a circular DNA into the host chromosome while causing duplication of the gene where the recombination event occurs. Two recombination events that occur within a single gene can exchange portions of the host chromosomal gene with the corresponding gene present on the added DNA.

Introduced DNA can also be stably integrated into host chromosomal DNA by means of a process that is not well understood. When a high concentration of DNA is used in the transformation procedure, the entering DNA appears to integrate in multiple copies at random sites in the chromosomal DNA. Unlike homologous recombination, this process does not appear to require extensive similarity between the introduced DNA and the site of chromosomal integration.

Detection of eukaryotic transformants

Just as the detection of bacterial transformants requires the use of a selectable marker, such as ampicillin resistance, and selective conditions that inhibit the growth of non-transformed bacteria, the detection of stable transformants in eukaryotic cells also requires the use of a selectable marker. Many of the selectable markers that are routinely used with bacterial transformations do not function well in eukaryotic cells, but a variety of selectable markers have been developed for use in eukaryotes. Some markers function by restoring, or complementing, a metabolic deficiency in the cells. The herpes simplex virus thymidine kinase (tk) gene has been used to detect transformants of mammalian cells deficient in thymidine kinase and the *E. coli* xanthine-guanine phosphoribosyltransferase gene (XGPRT) to detect transformants of cells deficient in the mammalian enzyme hypoxanthine-guanine phosphoribosyltransferase.

One disadvantage of selective markers like tk and HGPRT is that the cells transformed with these genes must initially be deficient in the corresponding enzyme activity (tk^- or $HGPRT^-$). Specific mutant strains or types of cells must be constructed and used as recipients in all DNA transformation experiments that use these selective markers. Selective markers that allow transformants to survive in the presence of a compound that kills normal, non-transformed cells can be used with a wider variety of recipient cells. The drug methotrexate prevents the growth of wild-type eukaryotic cells by inhibiting the enzyme dihydrofolate reductase (DHFR). Transformants can be easily detected by introducing multiple copies of a normal DHFR gene or by introducing a copy of the bacterial DHFR gene, which produces an enzyme that is resistant to methotrexate.

Other selective schemes use addition of an aminoglycoside protein synthesis inhibitor like kanamycin, neomycin, or G418 to inihibit the growth of non-transformed cells and select for transformants containing a neomycin phosphotransferase gene. These approaches allow transformation of a wide variety of recipient eukaryotic cell types without the need to construct specific mutant recipient cell lines prior to DNA transformation experiments.

Transformation of plants

Transformation of plants is a more complex task than transformation of animal cells. While a mammalian cell can be viewed as a membrane-bounded bag of cytoplasm, a plant cell should be considered more like a membrane-bounded bag of cytoplasm that has been placed inside a closed cardboard box. When maintained in a tissue culture system and grown as protoplasts that are deficient in cell walls, plant cells can be transformed using procedures similar to animal cell transformation. A normal plant cell, however, is surrounded by a rigid cellulose wall that interferes with the introduction of DNA by transformation methods that work with other eukaryotes. Several novel approaches to transformation of plants have been developed to circumvent this problem.

Crown gall disease, the formation of a gall or lump of undifferentiated plant tissue, is caused when pathogenic strains of the bacterium *Agrobacterium tumefaciens* infect a wound on a suseptible host plant (Figure 7.7). The formation of the gall of plant tissue is caused by the transfer of a small segment of DNA, the T-DNA, from a tumor-inducing or Ti plasmid that is normally present in the bacteria, to the plant cell. Integration of the T-DNA in the plant chromosome and expression of T-DNA genes causes alterations in the synthesis of the plant hormones that

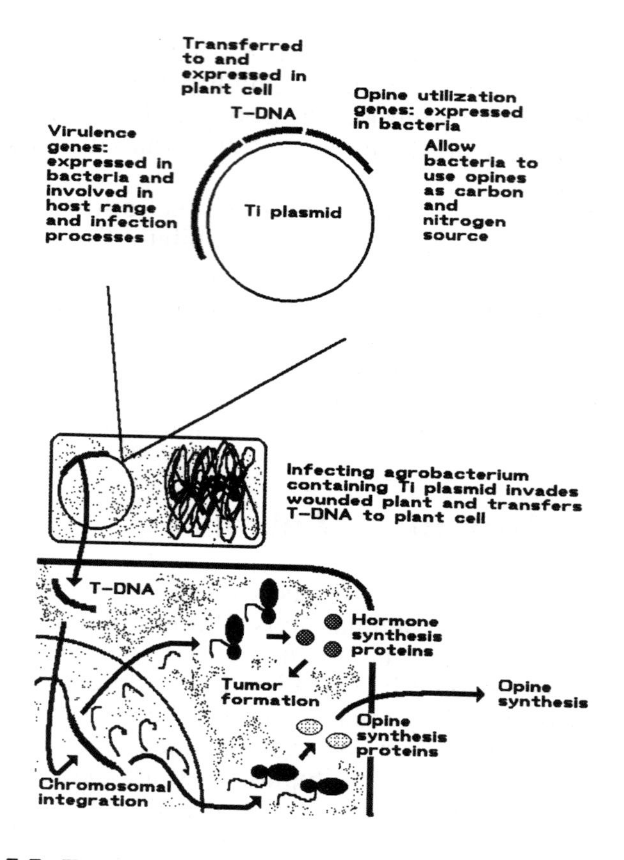

Figure 7.7. Transformation of plant cells by *Agrobacterium tumefaciens.* The infectious bacteria contain a large Ti (tumor- inducing) plasmid that contains three regions of interest to the infection process: the virulence region, the T-DNA, and a region containing opine utilization genes. The virulence region contains genes involved in the determination of host range and in the mobilization and transfer of the T-DNA into the plant cell. Once the T-DNA segment has been transferred to the plant, it moves to the nucleus where it is incorporated in the chromosomal DNA. Genes present in the incorporated T-DNA are then expressed in the plant cell. One set of genes produces proteins that synthesize plant growth regulators that cause the formation of a tumor or gall. Other T-DNA genes result in the synthesis of proteins that cause the plant cell to manufacture opines, modified amino acid derivatives. The opine utilization genes present on the Ti plasmid allow the infecting bacteria to utilize these opines as carbon and nitrogen sources. The infecting bacteria transfer DNA to the plant cell and transform the cell to produce a gall, a unique environmental niche that can be preferentially colonized only by bacteria that can utilize opines.

regulate plant cell division. As the gall develops, the expression of other T-DNA genes causes the gall to synthesize unusual organic compounds called opines that are then used as an energy source by the *Agrobacteria* that reside in the gall.

As the details of this infection process became more clearly understood, researchers realized that this was a natural case of genetic engineering of plants by bacteria and that the T-DNA transfer mechanism might be modified to allow the introduction of other DNA into plant cells. Subsequent isolation and *in vitro* manipulation of the T-DNA has led to the development of disarmed and binary vectors that allow the transformation of plant cells (Figure 7.8). The DNA transfer process mediated by the Ti plasmid enables efficient transfer of the recombinant vector to a host plant cell, where relatively random integration of the DNA into the plant chromosome can result in stable transformation of the cells.

One of the problems associated with the use of *Agrobacterium* as the basis for plant transformation systems is the host range of various *Agrobacterium* strains. While some *Agrobacteria* can infect a wide variety of plant hosts, infection is generally limited to dicotyledonous hosts (woody plants). Many of the agriculturally important plant species, including most cereals and grains, are monocots and cannot be readily infected by *Agrobacterium*. Plant viruses represent another potential source of vectors for transformation of plants. Viruses have a number of distinct advantages as cloning vectors for transformation of plant cells. Viruses that infect monocots are common, viruses are often present in very high amounts in infected tissues, and many viruses infect the host systemically, or throughout the entire plant.

Distinct problems also occur with the use of viruses as plant cloning vectors. The majority of plant viruses have genomes composed of RNA rather than DNA, and hence cannot be directly manipulated with restriction enzymes. The genomes of plant viruses can be multi-component with the different components packaged into separate particles that must be mixed together to obtain infection. Plant viruses often contain few regions that can be excised and replaced

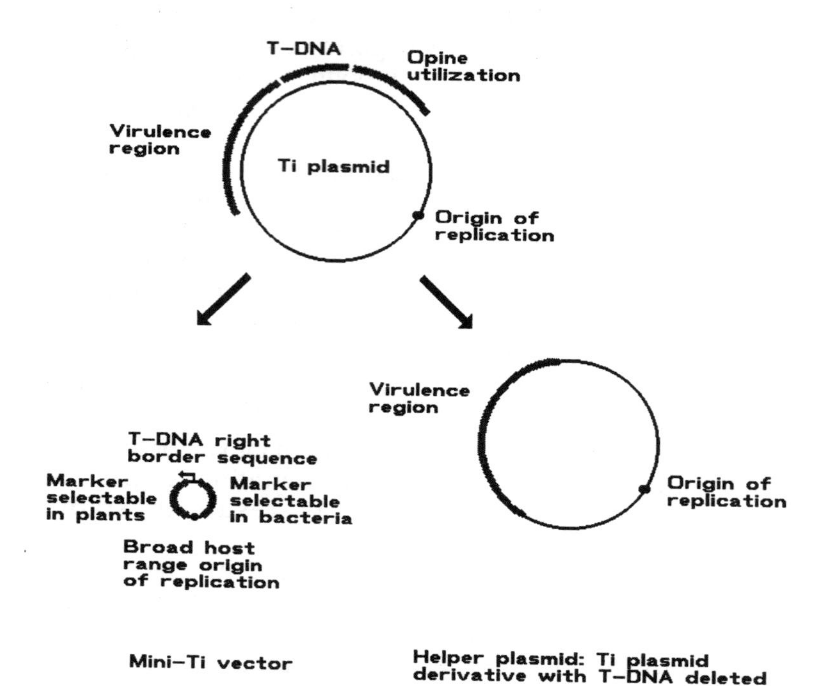

Figure 7.8. The ability of the Ti plasmid to transfer DNA into a plant cell has been used to facilitate transformation of plants with DNA cloning vectors. The undesireable aspects of the T-DNA (tumor formation and opine synthesis) have been discarded and the ability to the T-DNA to initiate DNA transfer from the right border of the T-DNA has been retained in mini-Ti vectors. In addition to the T-DNA right border, mini-Ti vectors generally also contain a marker selectable in bacteria (generally antibiotic resistance) and a marker that can be expressed from a plant RNA synthesis initiation region and detected in plants (generally resistance to kanamycin). Transfer from bacteria to a plant cell is promoted by a helper plasmid, a modified Ti plasmid that contains the virulence region but from which the T-DNA has been removed. When both the vector and the helper are present in a bacterium, the helper plasmid can promote transfer of the mini-Ti plasmid to the plant cell. Following integration of the mini-Ti in the plant chromosome, transformed plant cells can be detected by virtue of expression of the mini-Ti selectable marker (typically resistance to a kanamycin derivative). Note the conceptual similarities to the binary vectors used to promote transfer of DNA from *E. coli* to non-transformable bacteria (Figure 7.4).

with extra DNA, and the requirement for packaging of the replicated viral genome in an empty viral head or capsid may severely limit the amount of extra DNA that can be inserted into a viral genome. In spite of the complications, some plant viruses are currently being investigated as vectors for transformation of plants.

A number of mechanical methods have also been successfully applied to the transformation of plants. Electroporation, the use of a high-voltage electric field to drive DNA into cells, has been applied to transformation of protopasts and pollen, for example. Perhaps the most unusual method for delivering DNA into plant cells involves the use of a gun that fires small particles coated with DNA.

Just as it is somewhat more difficult to get DNA into plant cells than into many prokaryotes or other eukaryotes, detection of the transformed plant cells is also more tedious. The transforming DNA often includes the neomycin phosphotransferase gene and the transformed cell population must be grown in the presence of the protein synthesis inhibitor G418 to kill the non-transformed cells and select for the G418-resistant transformants. These selection schemes often require growth of the cells in complex tissue culture media and the formation of viable plantlets from the small lumps of undifferentiated transformed tissue can be tedious and time-consuming.

Although the methods used for introducing DNA into plant cells are often more complex and the selection of transformants more tedious than for other organisms, methods exist that allow the transformation of an increasing variety of plant species. As progress is made in understanding plant metabolism, plant transformations will undoubtedly become more routine.

The goal of transformation is to introduce DNA into an organism, sometimes to produce large amounts of the DNA for other experiments, sometimes to study the mechanisms that regulate expression of the genetic information present on the DNA, and sometimes to allow production of a desired gene product. For each different type of host cell used, a different transformation protocol may be required to introduce DNA into the cells. Once DNA is introduced, it must be allowed to replicate before various selection methods can be used to identify transformants. The recombinant DNA molecules present in the transformants must be recovered and further characterized by one or more of a variety of methods that are used to search or screen for a desired recombinant molecule.

Summary

1. **DNA must be introduced into a host cell to allow replication of the recombinant molecules.**

2. **Various chemical, mechanical, and electrical treatments can be used to make host cells competent to take up DNA.**

3. **Vectors derived from bacteriophage lambda DNA can be placed in empty bacteriophage heads and used to transfect host cells, a more efficient process than transformation.**

4. **Transformation procedures exist for both prokaryotes and eukaryotes, including many types of plant and animal cells.**

8

Identifying Recombinant Molecules

"'And who are **these***?' said the Queen, pointing to the three gardeners who were lying round the rose-tree; for, you see, as they were lying on their faces, and the pattern on their backs was the same as the rest of the pack, she could not tell whether they were gardeners, or soldiers, or courtiers, or three of her own children."* — Alice's Adventures in Wonderland, Lewis Carroll

Following transformation of competent cells with a ligated DNA sample, transformant colonies or plaques are obtained that must be characterized to identify the transformant that contains the DNA fragment of interest. Each of the transformants generally looks just like every other transformant and the transformant that contains a specific DNA sequence often cannot readily be identified in the population of transformants.

DNA libraries

In a typical cloning experiment, one of the principle goals is to construct sufficient numbers of recombinant molecules in a vector so that there is a significant probability that every gene is present in at least one recombinant molecule in the population. A collection of recombinant molecules

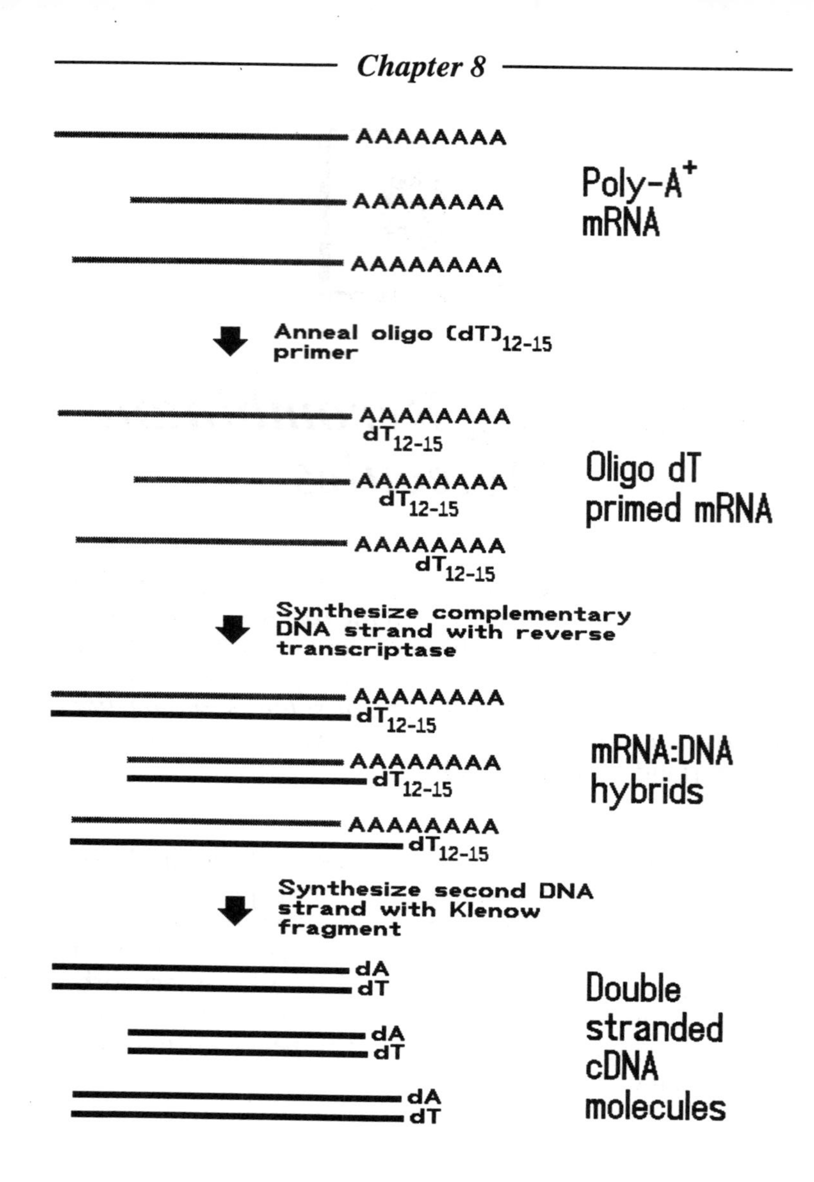

Figure 8.1. Synthesis of complementary cDNA from mRNA. Although RNA molecules cannot be directly cloned, it is possible to clone a cDNA copy of the mRNA molecule. Synthesis of cDNA from eukaryotic mRNA generally begins with the isolation of poly-adenylated mRNA. The stretch of adenosine residues that is present at the 3' terminus of eukaryotic mRNA molecules allows purification of the mRNA on oligo-dT columns that selectively bind the 3' poly-A residues. The poly-A+ mRNA is allowed to anneal with oligo $(dT)_{12-15}$, a stretch of dT residues

containing DNA from a specific organism is often referred to as a gene bank or a library. The probability that a specific gene is present in a library is related to the size of the gene, the average size of the inserted DNA in the recombinant vector molecules, and the total number of recombinants.

Two types of libraries are commonly encountered - genomic and cDNA. Genomic libraries are constructed from purified or partially fractionated chromosomal DNA and generally contain all sequences present in the chromosome, whether these sequences are part of active genes or simply inactive "spacer" DNA. cDNA libraries are made not from DNA, but from mRNA, the molecule that is synthesized only from genes that are actively being expressed at the time the RNA is isolated from the cell. RNA that has been isolated from a cell is used as a template to synthesize a complementary single-stranded cDNA molecule using the enzyme reverse transcriptase (Figure 8.1). The single-stranded DNA is then converted to a double-stranded DNA, often using Klenow fragment of DNA polymerase. The newly synthesized cDNA molecule is then ligated with a vector DNA to construct the cDNA library using the vectors and methods described in previous chapters. The cDNA library contains only DNA sequences corresponding to genes that were actively expressed as mRNA at the time the RNA was purified, and typically represents only a fraction of the sequences (<5%) present in a genomic library.

12 to 15 bases long to form a primed mRNA template. This primed template can then serve as a substrate for the enzyme reverse transcriptase, which begins DNA synthesis at the dT primer and synthesizes a DNA strand complementary to the RNA molecule. The product of reverse transcriptase is an mRNA:DNA hybrid and must be converted to a double-stranded DNA molecule prior to cloning procedures. The mRNA strand of the hybrid is removed and the second strand of the DNA molecule is synthesized with an enzyme like Klenow fragment of DNA polymerase I. The final product of the cDNA synthesis reaction is a double-stranded, blunt-end DNA molecule with one strand corresponding to and the other strand complementary to the original mRNA molecule. When the cDNA synthesis reaction is performed on a mixture of RNA molecules isolated from a specific tissue, the resulting cDNA sample will contain copies of all of the genes that were actively expressed at the time RNA was isolated from the tissue.

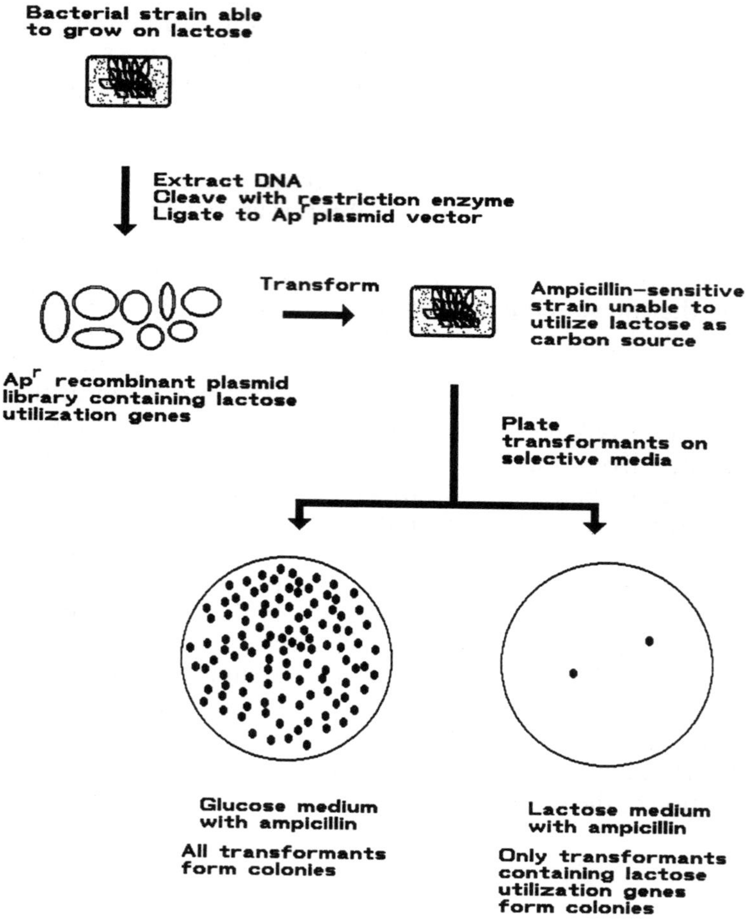

Bacterial strain able to grow on lactose

Extract DNA
Cleave with restriction enzyme
Ligate to Ap^r plasmid vector

Transform

Ampicillin-sensitive strain unable to utilize lactose as carbon source

Ap^r recombinant plasmid library containing lactose utilization genes

Plate transformants on selective media

Glucose medium with ampicillin

All transformants form colonies

Lactose medium with ampicillin

Only transformants containing lactose utilization genes form colonies

Figure 8.2. Positive selection for transformants containing lactose utilization genes. DNA isolated from a bacterial strain that is able to utilize lactose as a carbon source can be digested with a restriction enzyme and ligated into a vector that confers resistance to ampicillin to make a library containing the lactose utilization genes. This library can be transformed into an ampicillin-sensitive strain of bacteria that is unable to utilize lactose. When plated on medium containing glucose and ampicillin, all transformants will grow and form colonies. When plated on medium containing lactose and ampicillin, only transformants that contain the lactose utilization genes in a recombinant plasmid will be able to grow and form colonies.

Screening a library

The problem of searching through a library to find a desired cloned DNA fragment is affected significantly by the relative complexity, or number of different fragments present in the libary. In the simple case of a single DNA fragment ligated to a vector, every recombinant should contain the same DNA fragment and finding a recombinant containing the desired DNA insert is not difficult. A researcher could merely grow two of the transformants, extract small batches of DNA from each, and cleave the DNA with restriction enzymes to verify that the recombinant molecules contain the desired fragment pattern.

As the number of different inserts increases, the probability that a recombinant contains the desired insert will decrease. Consider the simultaneous ligation of five different DNA fragments to a vector. A transformant might contain any one of the five different fragments or a combination of more than one fragment. For a variety of reasons, the different fragments may not be present at the same frequencies in the transformant population. It may be necessary to search through many transformants before finding a recombinant containing the desired DNA fragment. With genomic or cDNA libraries that contain hundreds or thousands of different cloned DNA fragments, randomly searching through the population of transformants in the hopes of finding the desired DNA insert becomes an extremely inefficient approach.

Screening by positive selection

A variety of screening methods can help search through the recombinants. In the simplest circumstance, the desired gene will confer a phenotypic property, such as growth on a specific carbon source, and the desired recombinant molecule can be identified by positive selection (Figure 8.2). The genes that allow growth of *E. coli* on the sugar lactose (lac) as a carbon source might be identified by purifying DNA from a strain that is capable of growing on lactose

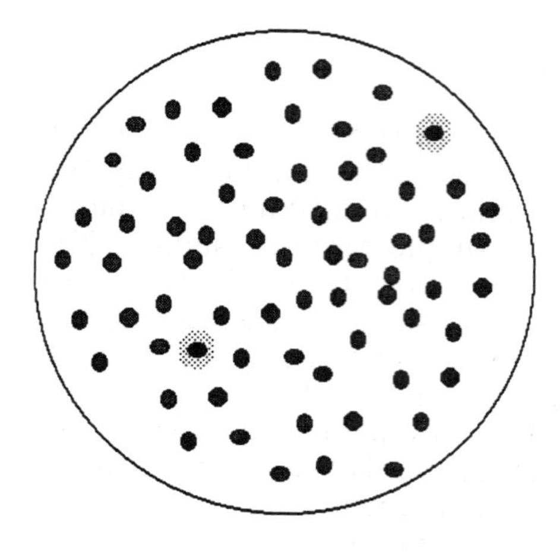

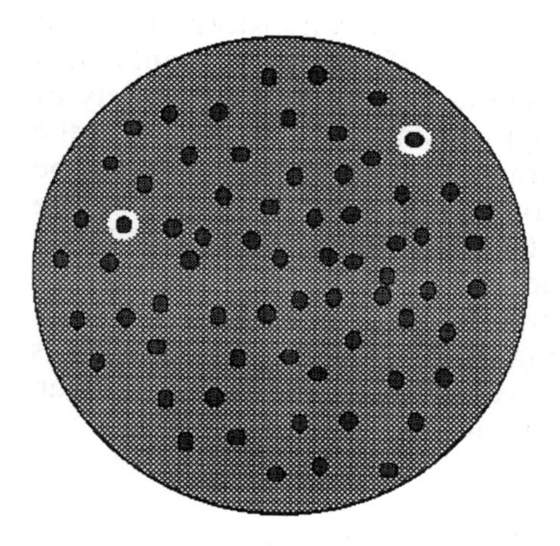

Figure 8.3. Detection of a new enzymatic property caused by the presence of a recombinant molecule containing a desired gene. Transformants can be plated on a medium that contains a substrate and an indicator dye that indicate the presence of the desired gene by the formation of color around a colony. For some gene products, such as enzymes involved in starch hydrolysis, a substrate added to the medium will be broken down only around colonies that contained the desired cloned genes. Note that in these screening methods, many colonies will grow on the plates but only the few colonies containing the desired genes will produce a characteristic reaction in the indicator medium.

(lac$^+$), ligating this DNA to a plasmid vector that can be selected with antibiotics, and transforming the ligation mix into a bacterial strain that cannot grow on lactose. Plating the transformants on medium containing lactose as a sole carbon source and an antibiotic to select for transformants that contain the plasmid vector will select for the growth of transformants that contain the desired lactose utilization genes.

Phenotypic screening

Because prokaryotes and eukaryotes do not have the same gene structure or genetic regulation systems, the cloned genes present in recombinant plasmids will not always be expressed in a transformed host. Because expression of eukaryotic genes in prokaryotic hosts is uncommon, screening by positive selection is rarely used for characterization of bacterial transformants containing eukaryotic genes. A more common method of finding a desired recombinant involves the use of a phenotypic screen that searches for a new property conferred on the bacteria containing the desired genes.

Several different types of property can be used to help identify cells containing the desired recombinant. The desired gene will generally give transformants a new phenotypic property that cannot be used to selectively enhance for the growth of, but that can be used to identify the desired transformant. It is necessary to screen all of the transformants for the desired property. These screens fall into three basic categories based on production of an active protein gene product, production of a gene product that is not necessarily active but can be detected with antibodies, or on the ability of the desired gene to hybridize with a nucleotide sequence that can serve as a probe. Because these screening methods are technically complex, but conceptually rather simple, a conceptual description of each approach will be given without technical details.

Production of an active gene product

The production of an active gene product can confer a new enzymatic property on transformants containing the desired gene (Figure 8.3). By subjecting each of the transformants to a test for this activity, potential recombinants of interest can be identified. For example, in cloning the genes involved in starch breakdown, it is possible to plate 500 transformants on a nutrient plate to allow growth of the cells, then overlay the colonies with a starch solution. Colonies that contain recombinant molecules that encode an enzyme involved in starch breakdown will often produce sufficient amounts of active protein that the enzyme will hydrolyze the starch around the desired colony to produce a halo or hole in the starch layer.

Screening with antibodies

Due to differences in gene structure and regulatory mechanisms, active protein products are not generally expected when eukaryotic genes are cloned in bacteria. Detection of proteins produced by recombinants frequently utilizes antibodies directed against the desired gene product (Figure 8.4). Antibodies react with determinants, or combinations of amino acid residues, that occur in proteins and activity of the protein is often not required for antibody recognition. In screening methods that use antibodies, antibodies against the desired protein are allowed to react with extracts from colonies of interest, and then the antibody-protein complexes are visualized with other reagents.

Because the entire protein need not be present for recognition by an antibody, specialized expression vectors have been developed that simplify the use of antibodies in screening. These vectors are designed so that fragments of DNA are inserted into a restriction site in a marker gene (often ß-galactosidase) that is efficiently expressed in bacteria. The inserted DNA fragments cause the production of fusion proteins composed of most of the marker gene

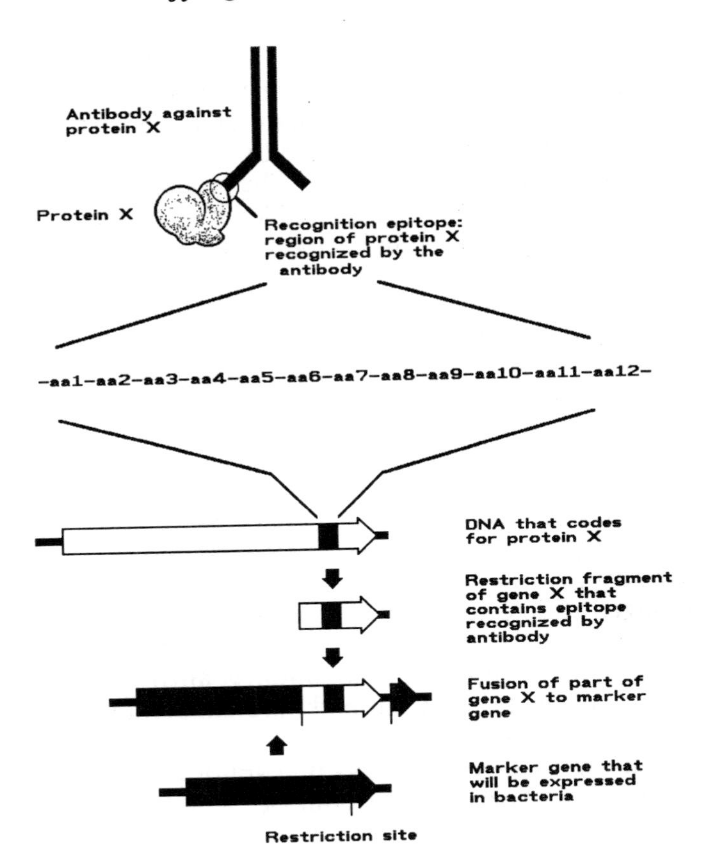

Figure 8.4. Hybrid genes can produce fusion proteins recognized by antibodies. Antibodies against a protein recognize a specific sequence or combination of amino acid residues. An antibody directed against protein X may recognize a 12 amino acid sequence of protein X. Although gene X may not normally be expressed in bacteria, the DNA sequence that codes for this region of protein X can be cut out of gene X with restriction enzymes and inserted into a marker gene (such as ß-galactosidase) that is efficiently expressed in bacteria. The recombinant DNA molecule is then a hybrid gene that consists of a portion of the marker gene fused with or attached to a portion of gene X. The region of gene X that codes for the amino acid sequence recognized by the antibody is now attached to a gene that can be expressed in bacteria. Expression of this hybrid gene in bacteria will cause the production of a fusion protein consisting of a portion of ß-galactosidase fused to a portion of protein X. The protein X sequences in this fusion protein can often be recognized by antibodies against protein X.

product fused to a stretch of amino acids that are encoded by the cloned DNA insert (Figure 8.4). Different DNA inserts will cause the production of different fusion proteins, some of which may be recognized by antibodies against a desired gene product (Figure 8.5). The production of fusion proteins ensures that proteins encoded by the cloned DNA inserts will be produced in bacteria and made available to antibodies during testing.

Detection of the antibody:protein complex is generally accomplished by using a reagent that detects the bound primary antibody (Figure 8.6). Primary antibodies are often detected with secondary antibodies directed against the primary antibody that have been conjugated with an enzyme like horseradish peroxidase. When the secondary antibody binds to the primary antibody, the enzyme that is linked to the secondary antibody becomes attached to the primary antibody:protein complex. The enzyme-secondary antibody:primary antibody:protein complex can be visualized or detected by adding a substrate for the enzyme and a dye that changes color during the enzymatic reaction. The enzyme-conjugated secondary antibody allows detection of the primary antibody:protein complex.

Immunodetection procedures can be performed on plates using antibody overlays in a gel or after transfer of proteins to a sheet of nitrocellulose or nylon membrane (Figure 8.7). Bacterial colonies or plaques containing fusion proteins can be bound to a nitrocellose filter and incubated with antibody against a desired protein. A detection reaction with a secondary antibody will identify cells containing recombinant molecules that produce proteins that are recognized by the primary antibodies. Colony blotting procedures that use antibodies as probes allow hundreds of transformants to be examined in search of a desired gene fragment.

Western blot analysis

While the colony blotting methods will work with the crude spots of proteins released from colonies or plaques, it is sometimes necessary to examine proteins in a manner that

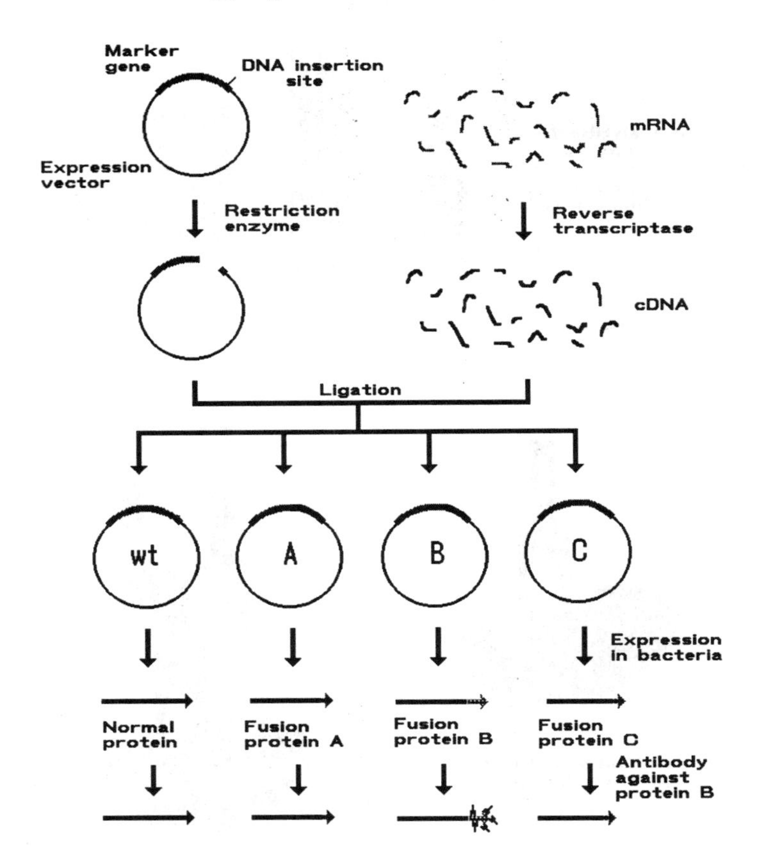

Figure 8.5. Reaction of antibodies with fusion proteins produced by expression vectors. Expression vectors that produce fusion proteins allow antibodies against a purified protein to be used to identify recombinants that contain fragments of a desired gene. In a typical experiment, cDNA synthesized from an mRNA sample that is known to contain the desired gene sequences is ligated into an expression vector. The recombinant molecules are transformed into a bacterial host and the transformants are induced to synthesize fusion proteins that contain a stretch of amino acid sequence derived from the cloned cDNA fragments. Each different recombinant molecule will produce a different fusion protein, depending on the cloned cDNA insert. An antibody against a specific protein will bind only to fusion proteins that contain amino acid sequences derived from the gene that encodes that protein. In the illustration, the antibody does not react with the normal marker gene protein or the fusion proteins A and C, but does bind to fusion protein B. This suggests that recombinant B may contain a fragment of the desired gene.

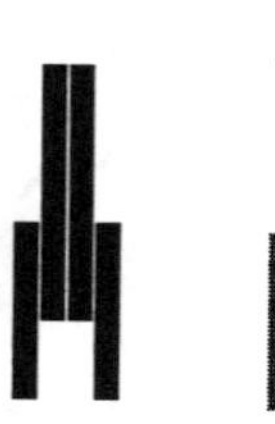

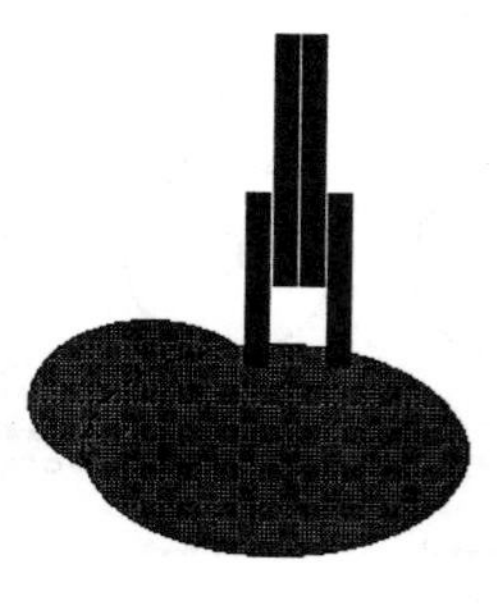

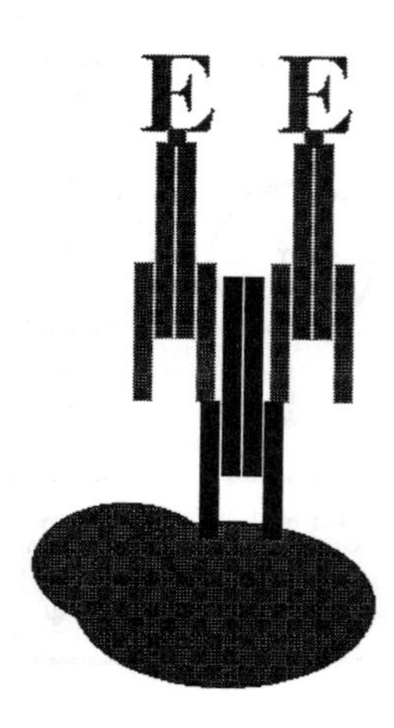

Figure 8.6. Antibodies bound to a protein (the primary antibodies) are detected with a secondary reaction involving a reagent that binds to the antibody and allows the antibody to be visualized. An antibody directed against the primary antibody and that has an attached enzyme molecule to allow visualization of the complex is frequently used in the secondary reaction. A rabbit antibody against human actin can be detected by a secondary reaction with a goat antibody against rabbit antibody. The presence of an enzyme like horseradish peroxidase on the secondary antibody allows detection of the secondary antibody:primary antibody:protein complex by incubation in a horseradish peroxidase reaction mix, which produces a colored spot wherever the complex occurs.

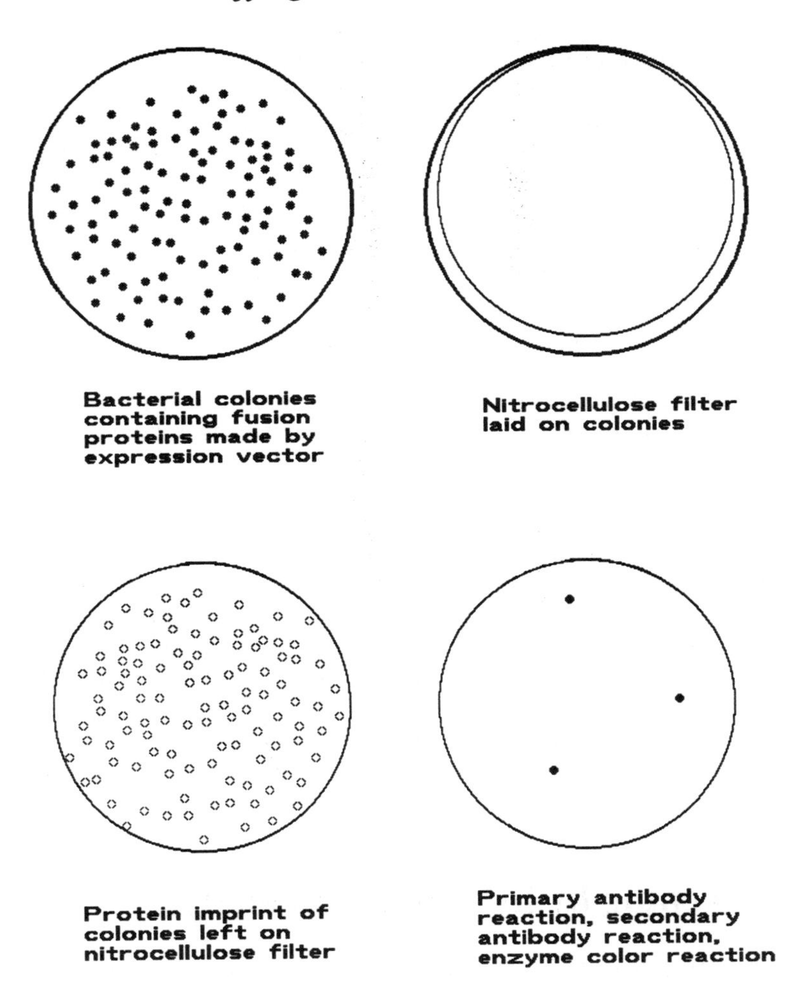

Figure 8.7. Use of antibodies to screen an expression library. The proteins produced by bacterial transformants are bound to a nitrocellulose filter disc to make a protein imprint of the colonies or plaques corresponding to the transformants. This filter is incubated first with a primary antibody against a specific protein, then with a secondary antibody against the primary antibody, then exposed to the appropriate enzyme color reaction to visualize the antibody:protein complexes. Spots of color that appear identify transformants that contain recombinant DNA molecules that produce fusion protein recognized by the primary antibody. These recombinant molecules are likely to contain fragments of the desired gene.

gives information about protein size. The Western blot procedure (Figure 8.8) allows accurate determination of the size of the antibody-reactive protein. Proteins are extracted and separated according to size by electrophoresis on an SDS-polyacrylamide gel, electrophoretically transferred to a membrane, and the antibody reactions performed on the imprint of protein bands bound to the membrane. The secondary antibody and color reaction allow direct visualization of proteins that react with an antibody.

Western blot analysis can be used to examine protein that has been synthesized either *in vivo*, in the cell, or *in vitro*, from purified DNA in a transcription/translation system (Zubay or S30 for prokaryotic systems, wheat germ or rabbit reticulocyte system for eukaryotes). In the *in vitro* systems, the added recombinant DNA serves as a template for synthesis of RNA by the enzyme RNA polymerase and the RNA serves as template for the assembly of proteins by ribosomes. If radioactive amino acids are included in the reaction, the newly synthesized proteins will be radioactive and can be distinguished from other proteins present in the transcription/translation system (for example, ribosomal proteins). Antibodies can be used to precipitate proteins and the immuno-precipitated proteins examined by SDS-polyacrylamide gel electrophoresis to determine the sizes of the antibody-reactive protein products.

Screening by hybridization analysis

The preceding detection methods all depend on the availability of a high-quality antibody directed against a desired gene product and on the ability to produce at least a small stretch of protein from the cloned DNA fragments. These screening criteria cannot be applied to all genes. Fortunately, screening methods exist that do not require either protein production or antibody availability. These methods utilize the capability of single-stranded nucleic acids to base pair, or hybridize, with complementary nucleic acid sequences (Figure 8.9).

Double-stranded nucleic acids can be denatured by heat to give complementary single-stranded molecules. This

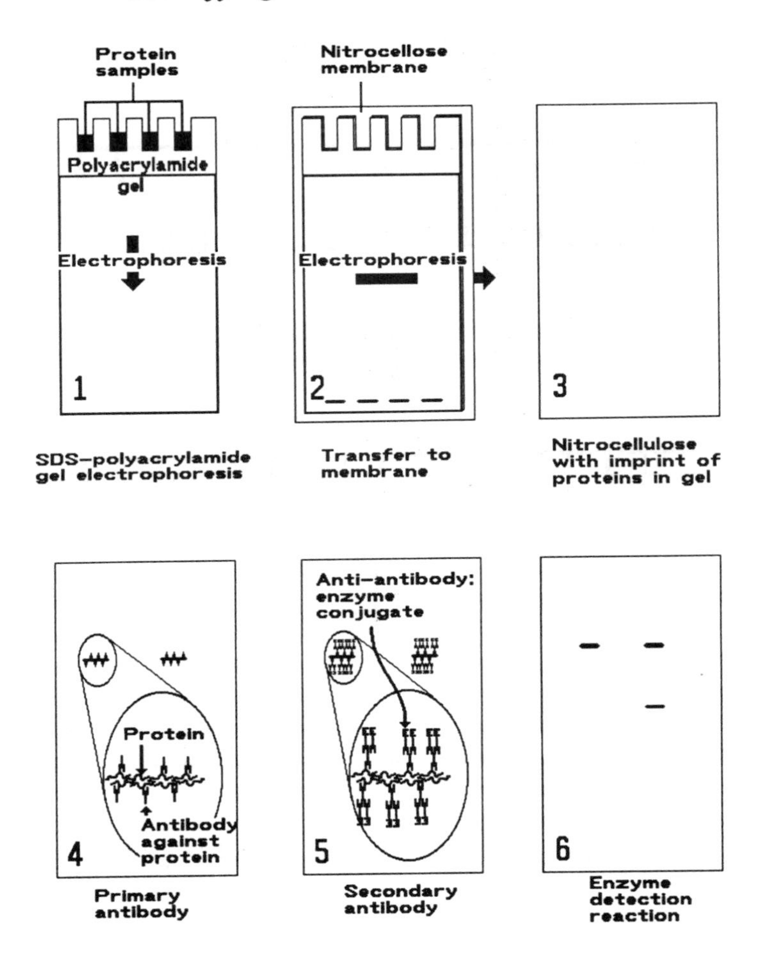

Figure 8.8. The first step in Western blot analysis involves (1) the use of SDS-polyacrylamide gel electrophoresis to separate proteins according to size. The gel is then placed against a sheet of nitrocellulose membrane (2) and electrophoresis is used to drive the polypeptide bands in the gel onto the nitrocellulose membrane. This produces (3) a nitrocellulose sheet with an imprint of the polypeptide bands that were present in the SDS-polyacrylamide gel. This imprint is incubated with (4) a primary antibody to form an antibody:protein complex with a protein of interest, then (5) incubated with a secondary antibody:enzyme conjugate to allow detection of the primary antibody. When placed in the enzyme reaction mix to visualize the enzyme conjugated to the secondary antibody (6), bands will appear at the location of antibody:protein complexes.

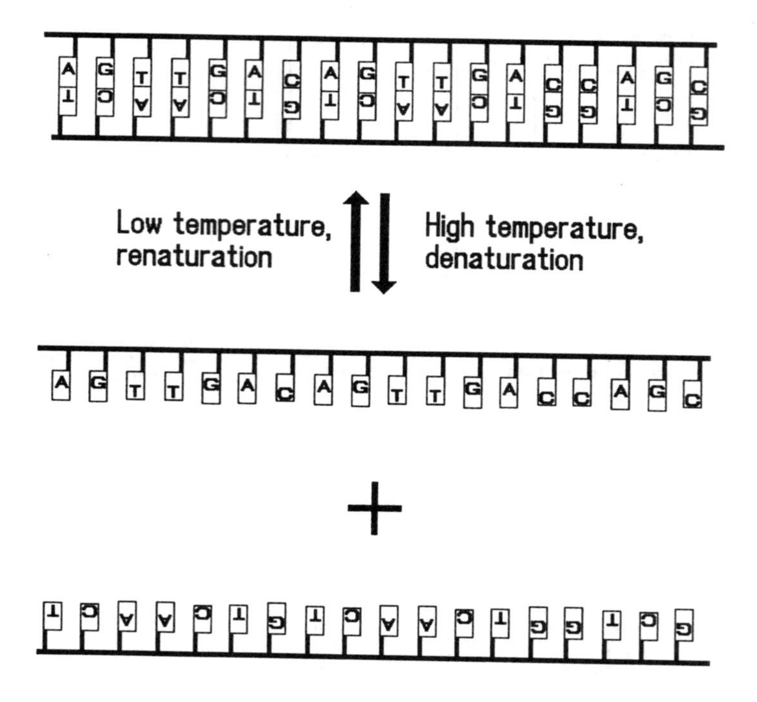

Figure 8.9. Principles of nucleic acid hybridization. Elevated temperature will cause a double-stranded nucleic acid to melt or denature to two complementary single-stranded molecules. When allowed to cool, these complementary molecules can reassociate, anneal, or renature to regenerate the double-stranded molecule. Several factors influence the tendency of nucleic acids to melt or anneal. High temperature, low salt concentration, hydrogen bond destabilizing reagents (such as formamide and urea), and high pH promote denaturation. The nucleotide sequence of a molecule also affects ability to form a duplex. Because a GC base pair forms three hydrogen bonds while an AT base pairs forms only two hydrogen bonds, a GC-rich sequence is more stable, or less likely denature at a given temperature, than an AT-rich sequence of the same size. Under conditions of high stringency, two sequences will anneal only if complementarity is 100%, while conditions of lower stringency allow the formation of duplexes with some degree of mis-matching.

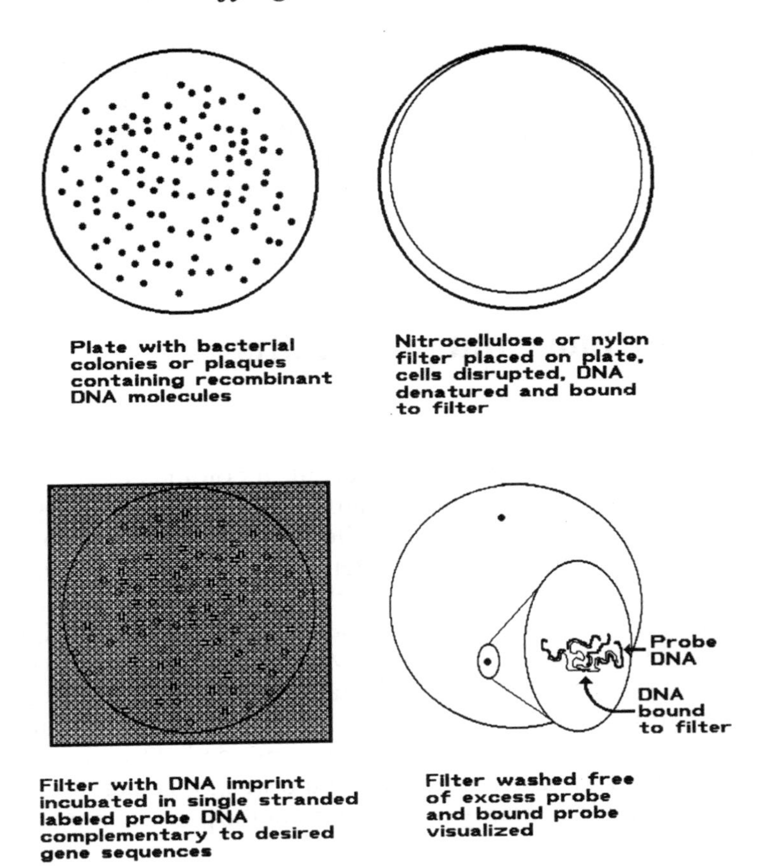

Figure 8.10. Colony blot and plaque lift hybridization. When a nucleic acid probe that is related to the gene of interest is already available, colony blot or plaque lift hybridization can be used to sort through large numbers of recombinant DNA molecules in search of a specific cloned gene. A library of recombinant DNA molecules is transformed into a bacterial host and colonies (for plasmid vectors) or plaques (for phage vectors) allowed to form on nutrient plates. A nitrocellulose or nylon membrane disc is then placed on the surface of the plate, the cells or bacteriophage are disrupted, the DNA is denatured, and a single-stranded DNA imprint of the recombinant molecules is bound to the membrane. The filter can then be incubated in a hybridization solution containing a denatured labeled nucleic acid probe. Under the appropriate conditions, the probe will anneal to complementary sequences bound to the membrane, allowing identification of colonies or plaques containing sequences complementary to the hybridization probe.

denaturation reaction is reversible and complementary single-stranded molecules can anneal or renature to form a double-stranded molecule. A nucleic acid sequence that codes for the desired gene product can be labeled (radioactive label, flourescent dyes, or enzyme conjugates) and used as a hybridization probe to anneal with and identify complementary sequences.

Colony blots and plaque lifts

The principles of nucleic acid hybridization can be used to detect the recombinant DNA molecules present in transformed cells. The DNA present in colonies or phage plaques can be denatured and transferred to a nitrocellulose membrane (Figure 8.10), then allowed to hybridize with a denatured nucleic acid probe. After non-specific background is washed away, the signal from the labeled probe that has annealed to complementary sequences bound to the membrane can be visualized. The degree of complementarity during hybridization need not be perfect, only sufficient to prevent the two strands from melting apart and separating during the washing steps needed to remove background. As the temperature of the washes is increased, greater complementarity is necessary for the probe to remain annealed.

The relative degree of mis-matching allowed during hybridization and washing is referred to as the stringency. Under conditions of low stringency, a hybridization probe can detect a sequence that may be only 50% identical, or homologous. Higher stringencies require a higher degree of similarity between the probe and the target sequences.

Nucleic acid hybridization has tremendous utility as a screening method. A nucleic acid probe related to the desired DNA sequence is required, but production of protein from the cloned DNA fragments is not required and the cloned genes need not be intact. As many as 30,000 recombinant transformants can be plated on the surface of a culture plate and simultaneously screened for the presence of a specific nucleic acid sequence. Specificity of the screening procedure can be adjusted to identify only

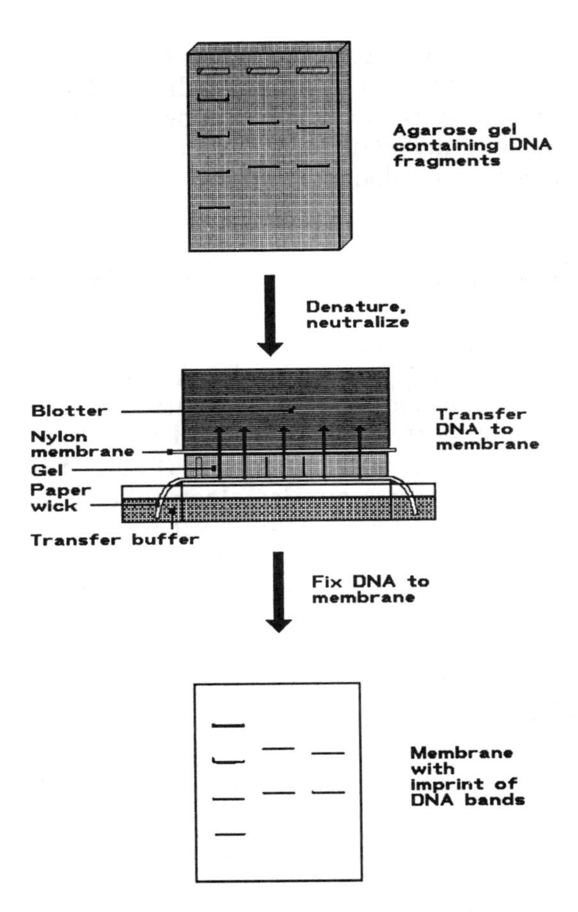

Figure 8.11. In Southern blot analysis a gel containing DNA fragments is soaked in solutions to denature, then neutralize the gel, converting the DNA fragments to denatured single-stranded molecules. The gel is then placed against a sheet of nitrocellulose or nylon membrane and the DNA fragments are transferred from the gel to the membrane. Transfer can be accomplished by electrophoresis or by simple capillary blotting. For capillary blotting, the gel is placed on a paper wick over a tray or transfer buffer. The membrane is placed on top of the gel, and paper towels placed on top of the membrane. The transfer soaks into the paper wick, through the gel, through the membrane, and into the paper towel blotter (as indicated by the arrows), carrying the DNA fragments out of the gel. As the DNA fragments are carried out of the gel, they are bound on the membrane, creating a single-stranded imprint of the DNA fragments originally present in the gel. This membrane can be hybridized with labeled nucleic acid probes to reveal DNA fragments that have homology with the hybridization probe.

recombinants with an extremely high level of similarity to the probe or to identify recombinants that contain sequences that are distantly related to the probe. Screening can be performed on a tiny bit of crude DNA released from a single bacterial colony or plaque.

Southern and Northern blot analysis

Hybridization techniques are also frequently used with both purified DNA and RNA samples. DNA can be digested with a restriction enzyme, the fragments transferred to a membrane support, then hybridized with a probe in a procedure known as a Southern blot (Figure 8.11). RNA can be subjected to a similar electrophoresis/transfer/hybridization procedure called a Northern blot. These two procedures are technically very similar, but DNA is bound to a membrane in a Southern blot and RNA is bound to a membrane in a Northern blot.

Hybrid-select translation

Nucleic acid hybridization and immunodetection have been combined in a procedure called hybrid-select translation. A DNA sequence that is suspected of containing a desired gene is allowed to hybridize with mRNA purified from cells that are known to make the desired protein product. The mRNA that anneals to the DNA sequence is then recovered and translated *in vitro*, and the translated proteins are exposed to antibodies against the desired product. Proteins that are precipitated with the antibodies are examined by gel electrophoresis to determine whether the DNA sequence has selected the correct mRNA from the mRNA preparation. This method can be used to help confirm that a recombinant molecule actually contains a fragment of the desired gene.

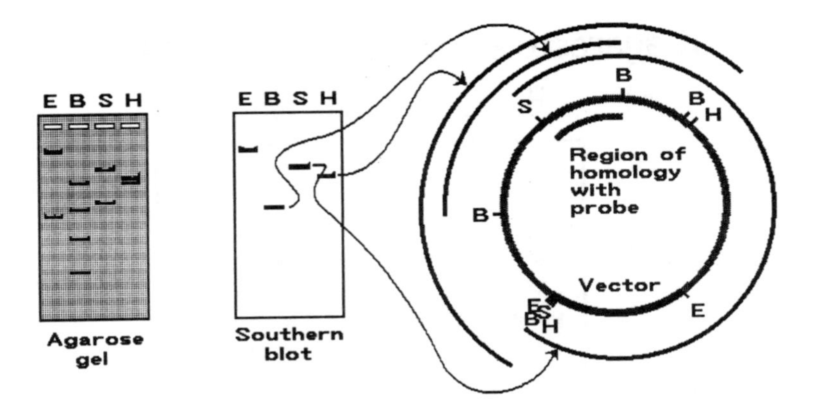

Figure 8.12. Restriction mapping combined with Southern blot analysis. The nucleotide probe that was used for identification of a recombinant molecule by hybridization analysis can also be used as a Southern blot hybridization probe. The agarose gel used to prepare the restriction map of the DNA molecule can be denatured, transferred, and hybridized with the probe. The DNA fragments that hybridize with the probe can then be positioned on the DNA restriction map to identify the minimum region of hybridization with the probe. In the illustration, a recombinant plasmid containing a large *Eco*RI DNA insert has been digested with *Eco*RI (E), *Bam*HI (B), *Sal*I (S), and *Hin*dIII (H) and hybridized with a cDNA probe. The fragments that hybridize with the probe have been indicated on the restriction map. The overlap between the *Bam*HI fragment and the *Sal*I fragment identifies the smallest region of homology with the probe.

Restriction mapping combined with hybridization

Following identification of a recombinant clone of interest, the additional DNA fragment(s) present in the vector must be characterized, usually by digesting the recombinant molecule with several restriction enzymes. The positions of cleavage sites in the DNA are determined relative to one another by simultaneous digestion with more than one enzyme to form a restriction map of the cloned DNA fragment. Preliminary characterization of this sort is often done with the small amounts of DNA obtained through miniscreen procedures.

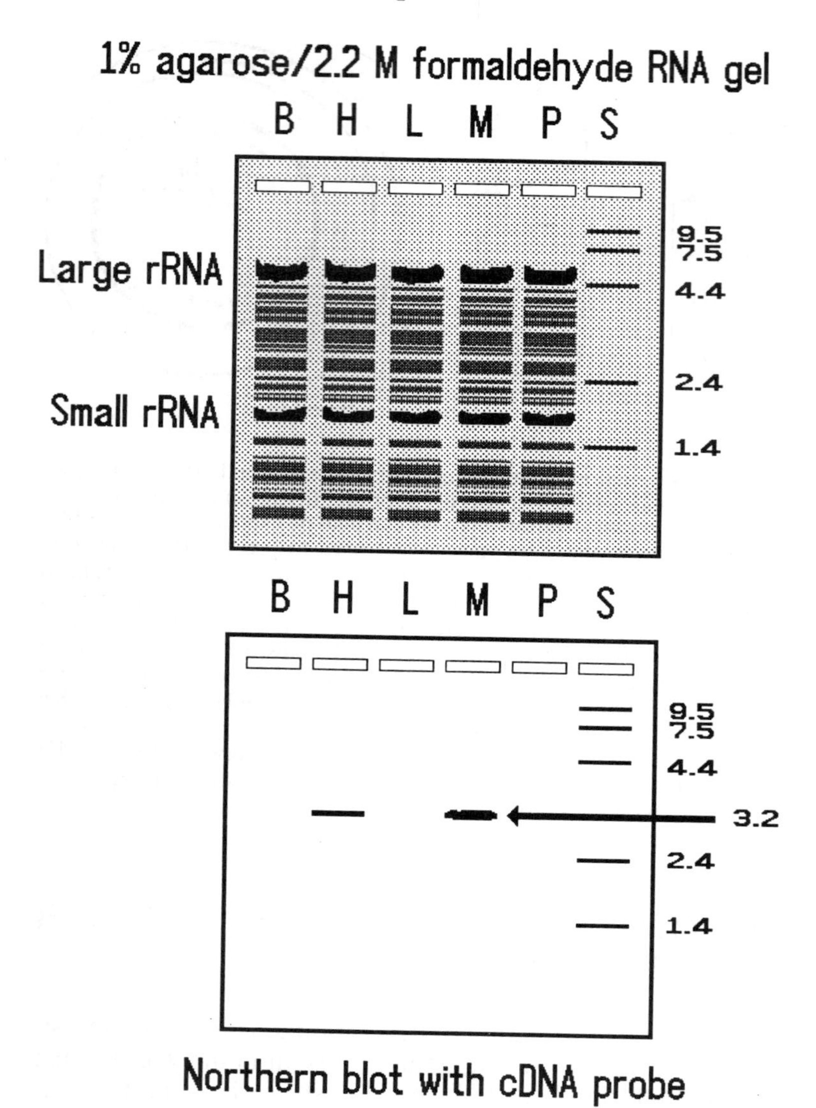

Figure 8.13. Northern blot analysis using a cloned DNA fragment as a hybridization probe to verify that the cDNA fragment has the properties expected of the desired gene. A cDNA fragment that contains only a portion of the gene of interest can be used as a hybridization probe against the RNA present on a Northern blot to verify that the cDNA fragment has properties consistent with the desired gene. In the example

Northern blot hybridization with cloned DNA fragments

Hybridization procedures are often used to further characterize or to help verify that a cloned DNA fragment contains the sequences of interest. For example, restriction mapping and Southern blot analysis can be combined to localize the precise regions of interest within a large cloned DNA fragment (Figure 8.12). A cloned DNA fragment can also be used as a hybridization probe against RNA in a blot analysis to verify that the cloned DNA fragment has the properties anticipated for the desired gene (Figure 8.13).

Following the construction of transformants containing recombinant molecules, the desired clone must be identified in a background of other transformants. A variety of screening methods have been developed to help identify and characterize recombinant DNA molecules. Application of these screening procedures to characterization of the cloned DNA fragments can help verify the identity of cloned DNA fragments and delineate precise regions of interest for further characterization.

shown, antibody against a 100 kDa muscle-specific protein was used to screen an expression library constructed from rat muscle to isolate a 200 base pair cDNA fragment. The cDNA fragment was used as a hybridization probe on a Northern blot containing 20 micrograms of RNA purified from rat brain (B), heart (H), liver (L), muscle (M), and pancreas (P). The sizes of RNA molecular weight standards (S) are indicated in kilobases at the right of the gel. Although the stained gel reveals prominent large and small rRNA species, few differences are obvious in the mRNA molecules, which are present in much lower levels. Hybridization of the blot with the cDNA probe, however, revealed that a 3,200 base RNA molecule present in heart and muscle was absent from other tissues. This demonstrates that the cDNA fragment has the specificity expected for the desired gene (only expressed in muscle tissues) and has sequence homology with an RNA molecule large enough to encode the 100 kDa protein (3,200 bases ÷ 3 bases per codon x 110 g per average amino acid residue in protein = 117 kDa potential protein).

Summary

1. **Transformants must be characterized and desired genes identified.**

2. **Positive selection requires that the cloned gene produce an active gene product that gives the transformant a growth advantage under certain conditions.**

3. **Antibodies against a protein are often used to identify recombinants. Expression vectors and gene fusions are used to produce hybrid fusion proteins for detection by antibodies.**

4. **Nucleic acid hybridization is used in a variety of forms to identify and characterize DNA molecules.**

Characterizing Genes

"Now the skillful workman is very careful indeed about what he takes into his brain-attic. He will have nothing but the tools which may help him in doing his work, but of these he has a large assortment, and all in the most perfect order."
— A Study in Scarlet, Sir Arthur Conan Doyle

Molecular biology, the study of the interactions of the macromolecules present in living cells, is a rapidly evolving field of biological investigation. The methods that allow the isolation and manipulation of DNA fragments, the tools of molecular biology researchers, developed out of scientific interest in testing important hypothesis regarding the structure and function of genes. Recombinant DNA methods provide extremely powerful tools for asking very detailed questions about the biological processes functioning in living cells. The current state of understanding of gene function is constantly changing as a result of the application of these methods to very complex biological questions. Analysis of gene structure and function can appear to be technically rather complicated, but most of the analytical approaches rely on the methods described in previous chapters.

Verifying gene identity

Once a DNA fragment that appears to contain a desired gene has been identified and isolated from a library, it is important to verify that the DNA fragment codes for the gene of interest. The method that is used to verify gene identity depends in some ways on how much previous information is known about the desired gene or gene product. If the same gene has already been isolated from another organism, comparison of the two nucleotide sequences may verify gene identity. If the corresponding protein has been purified and the amino acid sequence determined, verification of DNA fragment identity can be accomplished by determination of nucleotide sequence and conversion of the nucleotide sequence to a predicted amino acid sequence, which can then be compared with the actual amino acid sequence of the protein. When little sequence information exists about the desired gene or gene product, verification of DNA fragment identity may require genetic complementation analysis. Many of the methods used for verifying DNA fragment identity are also used for characterizing gene structure and function.

Mutagenesis of isolated DNA fragments

Mutagenesis of several types is often used to verify that the DNA fragment identified actually corresponds to the gene of interest or to localize specific regions of interest within a DNA fragment. Mutagenesis methods include the construction of insertions and deletions, the use of transposable elements, the incorporation of synthetic oligonucleotides, and the use of chemical mutagens. All these methods are designed to introduce mutations, changes in the nucleotide sequence of the DNA fragment, to determine the effects of the alterations on expression of the gene or function of the mutated gene product.

Construction of deletions or insertions

When a cloned DNA fragment can be shown to code for the desired gene product (for example, by complementation analysis), restriction enzymes are often used to construct deletions that remove portions of the cloned DNA fragment

(Figure 9.1). Deletions that prevent production of the functional gene product must remove at least part of the gene from the DNA fragment. Deletion analysis can be used to help identify the minimum region necessary for function of the cloned gene. Although deletions can be generated by a number of different methods, the specific cleavage properties of restriction enzymes facilitate the rapid generation of very specific deletions.

Restriction enzymes can also be used in the construction of insertions, extra DNA fragments inserted at various locations within the cloned DNA fragment (Figure 9.2). Just as deletions are used to remove portions of DNA to determine whether the deleted regions are important for gene function, insertions that occur within regions that are important for gene function can prevent production of the active gene product. Any DNA fragment could in principle be inserted, but it is most convenient to insert a DNA fragment that can be easily detected to minimize the screening necessary to find the desired insertion. Gene cartridges or cassettes are DNA fragments that contain an easily selected genetic marker, such as resistance to an antibiotic, and that have convenient restriction enzyme sites at their ends to facilitate insertion into restriction sites present in other DNA fragments. The gene cartridge can be inserted into a cloned DNA fragment, the insertion detected in the recombinant by virtue of the antibiotic resistance, then the insertion mutation examined for effects on expression of the cloned gene.

The construction of deletions and insertions can help verify that a recombinant contains the desired gene and aid in defining the regions of a cloned DNA fragment that correspond to the gene of interest. Although these methods may give less precise gene localization than other methods, results are easily obtained and help define regions for more detailed analysis.

Use of transposable elements

While the insertion of gene cartridges to map gene location can be extremely useful, it is dependent on the existence of

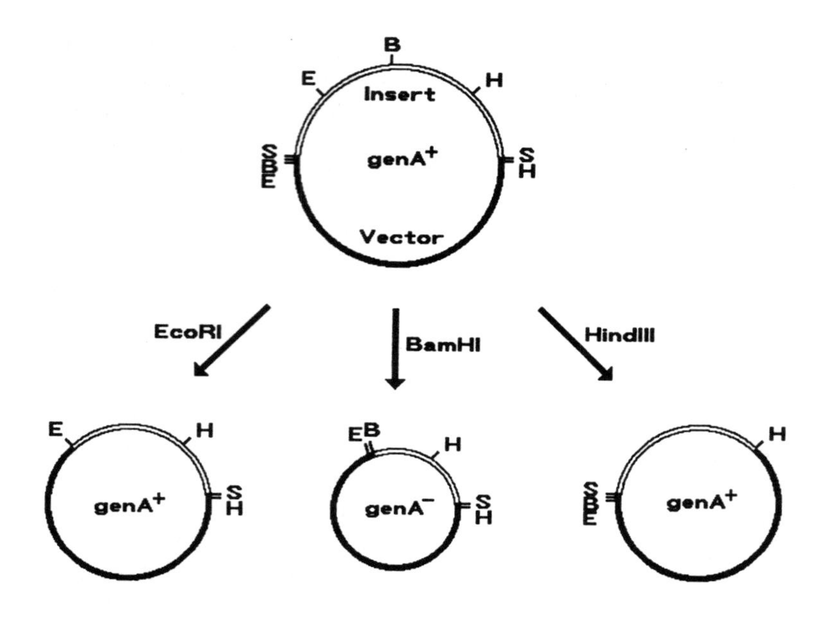

Results of deletion mapping

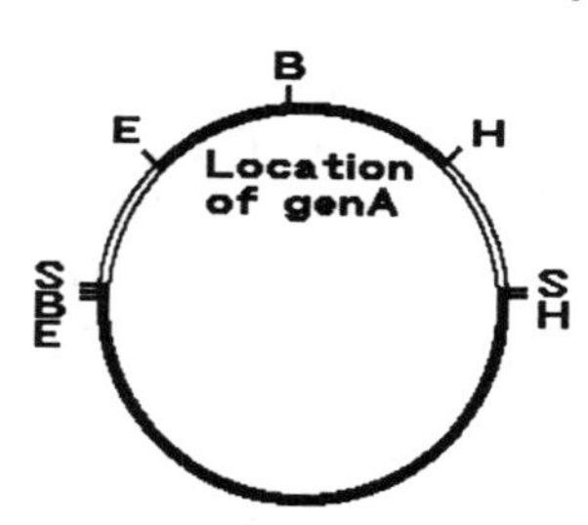

Figure 9.1. Deletion mapping used to characterize a cloned DNA fragment. A restriction map of a DNA fragment that confers a desired phenotype (genA$^+$) is first prepared. Restriction enzymes are then used to delete specific regions of the DNA fragment. The pnenotype of each deletion derivative is then determined. Those that retain the gene function (genA$^+$) must contain the intact gene (such as the *Eco*RI and *Hin*dIII deletion derivatives in the figure) and those that lose the gene function (genA$^-$) must have lost at least a portion of the gene (such as the *Bam*HI deletion derivative in the figure). By comparing the results with different deletions, it is possible to identify a smaller DNA fragment that contains the desired gene (such as the central region between the *Eco*RI and *Hin*dIII fragments in the figure).

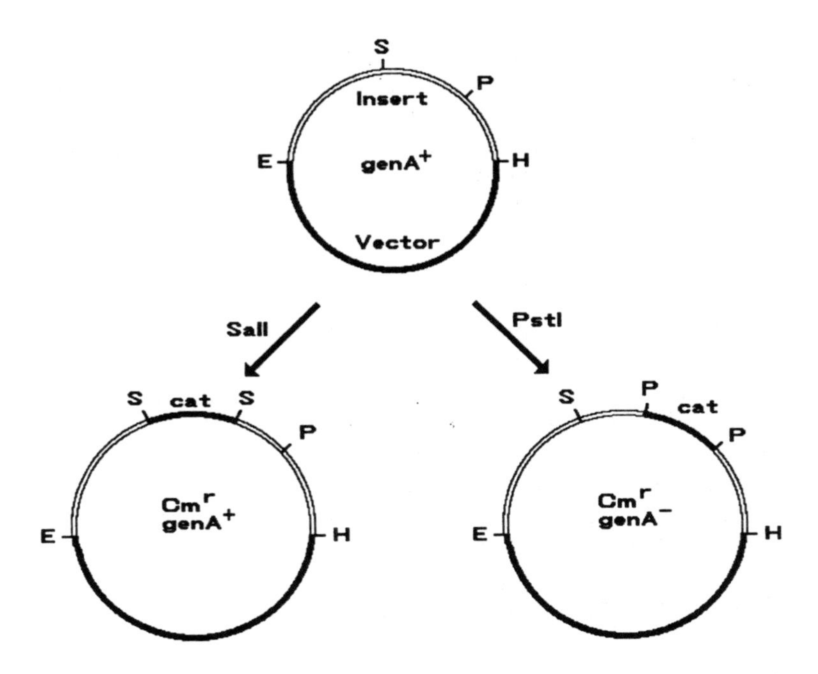

Results of insertion mapping:
PstI site must occur within genA

Figure 9.2. Insertion mapping used to characterize a cloned DNA fragment. A DNA fragment can be inserted into each of the unique restriction sites within a cloned DNA fragment. Although any DNA fragment could be inserted, DNA fragments that carry a gene for an easily detected phenotype are most convenient to work with. A gene cartridge that carries the gene for chloramphenicol acetyltransferase (*cat*), which confers resistance to chloramphenicol (Cm^r), could be inserted into the *Sal*I and *Pst*I sites of a cloned DNA insert that contains gene A ($genA^+$). The insertion derivatives could be selected by their ability to make host cells resistant to chloramphenicol (Cm^r), then checked to see if gene A remains functional. In the example shown, insertion of the *cat* gene into the *Sal*I site did not inactivate gene A ($genA^+$), while insertion into the *Pst*I site inactivated the gene ($genA^-$). Because the insertion inactivates genA, the *Pst*I site must occur within the region of the DNA insert that codes for this gene.

restriction sites within the regions of interest. A DNA fragment containing few restriction enzyme cleavage sites might appear to be resistant to insertion mapping procedures. To avoid the requirement for restriction cleavage sites, insertion mapping is often performed with transposable elements. Transposable elements or transposons are naturally occurring DNA sequences that are capable of residing in one DNA molecule and inserting a copy into another DNA molecule. Many well-characterized transposons (indicated by the prefix Tn) contain a gene that confers resistance to an antibiotic (Tn3, ampicillin; Tn5, kanamycin; Tn9, chloramphenicol; Tn10, tetracycline) that allows easy selection of cells containing the element. Although transposons may show preferential insertion into certain DNA sequences, insertion is generally much more frequent that the occurrence of restriction enzyme cleavage sites in a DNA fragment.

Transposon insertion mapping is frequently used to obtain more precise localization of a gene within a cloned DNA molecule (Figure 9.3). The principles of transposon mapping are fairly simple. A cell containing the recombinant of interest is transformed or infected with a second vector that carries the transposon. The transposon is allowed to insert in the target DNA, then DNA is extracted from the cells. This DNA is used to transform recipient cells and the transformants are grown under conditions that select for both the recombinant molecule and the transposon. The DNA from each transposon-containing transformant is extracted and the transposon insertion site is determined by restriction enzyme digestions. The phenotype of each transposon-containing recombinant molecule is also determined. Correlation of the transposon insertion site with the phenotype of the recombinant molecule allows establishment of a fairly precise genetic and physical map of the location of a gene.

Use of chemical mutagens

A variety of chemical compounds like nitrosoguanidine, hydroxylamine, nitrous acid, and intercalating agents are capable of causing changes in the nucleotide sequence of DNA. These sequence changes can occur either in regions of a DNA fragment that are critical to function of a gene or in regions that are not important (Figure 9.4). The changes that affect gene function can be recognized as mutations by the resulting production of different amounts of the gene product or by the production of an altered gene product, while changes that have no detectable effect are termed "silent" mutations and must be identified by nucleotide sequence analysis. Mutational analysis is extremely specific and allows investigation of the role of an individual nucleotide residue in the function of a gene.

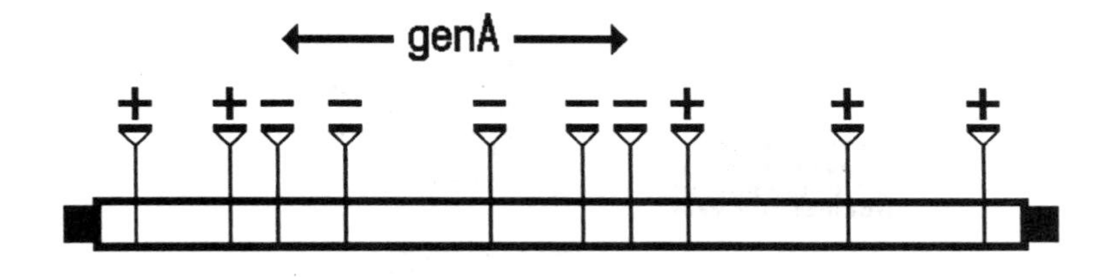

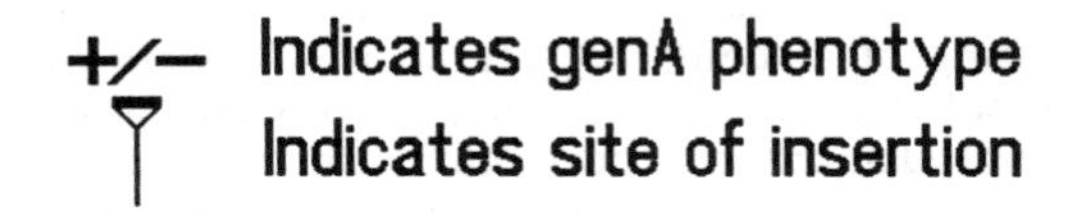

Figure 9.3. The use of transposable elements in insertion mapping. Transposable elements, mobile DNA sequences that can insert a copy of themselves from one DNA fragment into another fragment, can be used to generate a series of insertions in a cloned DNA fragment. Each insertion can be detected by screening for the phenotype associated with the transposon (Tn3, for example, confers resistance to ampicillin), then restriction enzymes used to map or determine the location of the insertion. Each insertion derivative is then examined for the function of the gene of interest (genA) and the genA phenotype is plotted against the locations of the insertions. Insertions that inactivate the genA phenotype identify the genA coding region. Correlation of the transposon insertion site with the phenotype of the recombinant molecule allows establishment of a fairly precise genetic and physical map of the location of a gene.

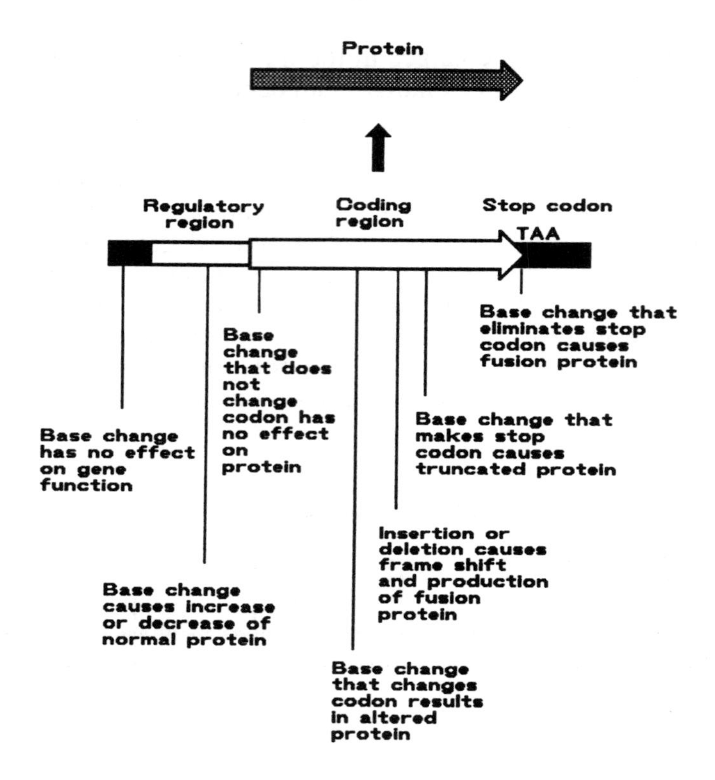

Figure 9.4. Types of single-base mutation affecting function of a simple gene consisting of a regulatory region involved in transcription initiation, a protein coding region, and a stop codon signifying the end of the protein. Mutations that occur within and around a gene can have very different effects depending on the location and the type of mutation. Mutations within the regulatory region can cause increased or decreased transcription, resulting in altered amounts of an otherwise normal gene product. A base change within the coding region can change a codon, resulting in the incorporation of a different amino acid (AAA [Lys] mutated to AAC [Asn]). Mutation could also generate a new stop codon that prematurely terminates the protein (AAA [Lys] mutated to TAA [Stop]). A base change in the normal stop codon could cause protein synthesis to continue past the stop codon, causing production of a protein larger than normal (TAA [Stop] mutated to AAA [Lys] allows production of a larger protein). Insertion or deletion of a base causes a change in the reading frame used to produce the protein, resulting in production of an altered fusion protein (AAATTTAAATTT [Lys PheLysPhe] mutated to AAATTAAATTT [LysLeuAsn]). Many mutations, including mutations in the non-coding region and mutations that do not change codons (AAA [Lys] mutated to AAG [Lys}), have no effect on gene function and are called "silent" mutations.

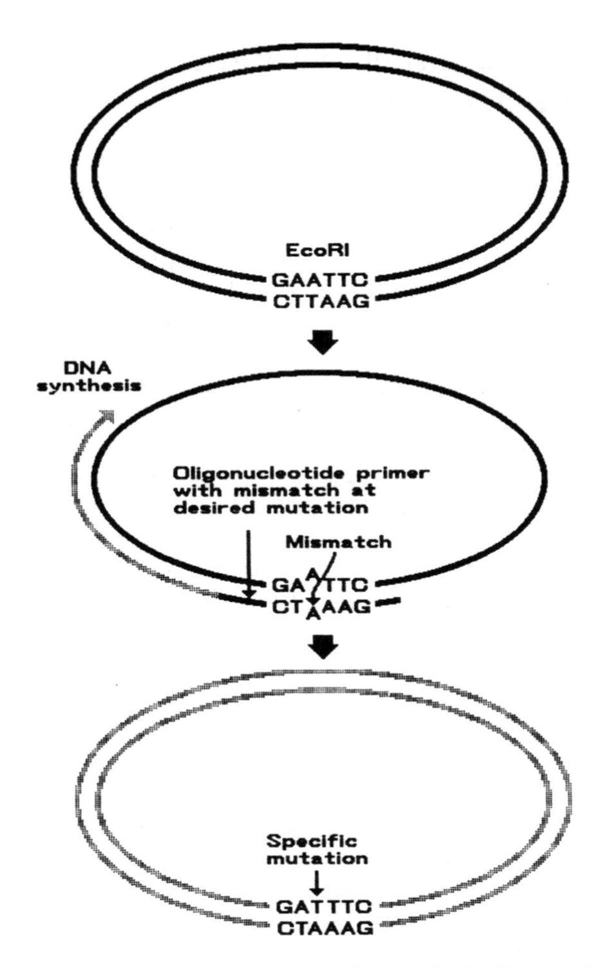

Figure 9.5. Site-directed mutagenesis using synthetic oligonucleotides. Once the nucleotide sequence of a DNA fragment has been determined, very specific mutations can be introduced into the DNA by first synthesizing a DNA fragment that contains the desired base change, then incorporating this oligonucleotide into the DNA to replace the original base with the desired mutation. In the example illustrated, the *Eco*RI site (GAATTC) is mutationally inactivated by synthesizing an oligonucleotide that includes a single A->T base change that converts the GAATTC sequence to GAAATC. The oligonucleotide containing the desired mutation is annealed to the original DNA with annealing conditions adjusted to allow for the mismatch at the site of the mutation. DNA polymerase can then be used to elongate the annealed primer and the newly synthesized strand recovered with the desired mutation present.

Incorporation of synthetic oligonucleotides

Chemical mutagenesis methods are usually fairly random and may simultaneously generate more than one mutation in a DNA molecule. Site-directed mutagenesis can be performed to modify a specific base within a gene without affecting the remainder of the nucleotide sequence. This is generally accomplished after the nucleotide sequence of a DNA fragment has been determined and the role of a particular region of the sequence has been established as important to gene function. Specific mutations can be constructed by synthesizing an oligonucleotide that contains a desired mutation and incorporating the oligonucleotide into the gene of interest (Figure 9.5). Directed mutagenesis is much more specific and efficient than random chemical mutagenesis and allows the investigation of the role of individual nucleotide residues in gene structure and function.

Determination of DNA sequence

Determination of nucleotide sequence is the basis for much of the research concerning gene structure and function. The sequence of the nucleotides in a DNA fragment constitutes the information stored in a gene. Since the genetic code has previously been determined, it is possible to convert the nucleotide sequence of a gene into a corresponding protein sequence by the use of computer programs that translate nucleotide sequence information into protein sequence. If the precise beginning and ending of the gene can be determined, then the exact sequence and size of the protein product of the gene can easily be calculated.

This of course assumes that genes are co-linear and are not interrupted by other sequences, which is unfortunately not the case for many genes, particularly for eukaryotic genes. It has been determined that many eukaryotic genes contain intervening sequences called introns that are present in the genomic DNA but are not present in the mRNA. These sequences are transcribed but are removed and the coding regions of the mRNA, called exons, are assembled into a mature message prior to translation. It may therefore be necessary to determine the sequences of both the genomic DNA and the mRNA (or the cDNA copy of the mRNA) to fully understand the structure of the gene.

Nucleotide sequence analysis

Two methods are frequently used for determination of the sequence of the nucleotide residues in DNA: the Maxam-Gilbert or chemical method and the Sanger or dideoxynucleotide method. In the chemical method (Figure 9.6), a DNA fragment is purified and a radioactive label, usually ^{32}P, is introduced at one end of the molecule. Aliquots of the DNA fragment are subjected to four (or more) different chemical reactions that preferentially cleave the DNA strand at specific bases. The reaction conditions are controlled so that each DNA fragment is cleaved once during each reaction. The DNA fragments are then denatured and subjected to electrophoresis in a polyacrylamide gel containing 7 molar urea to ensure that fragments remain denatured and migrate according to their respective sizes. When the gel is exposed to X-ray film, the cleaved, end-labeled DNA fragments will expose the film and can be visualized as bands on the film. Since the smallest bands migrate fastest, it is possible to compare the four chemical reactions and, beginning at the bottom of the gel, simply read the nucleotide sequence of the DNA fragment.

The Sanger or chain termination method (Figure 9.7) uses a single-stranded DNA template to which a short (generally 17-20 bases) oligonucleotide primer has been annealed. When the enzyme DNA polymerase (Klenow fragment of DNA polymerase I, reverse transcriptase, or T7 DNA polymerase) is added to the primed template in the presence of deoxynucleotide triphosphates, a DNA synthesis reaction will occur. If one labeled triphosphate, such as ^{32}P-dATP, is present in the reaction, the newly synthesized DNA will be radioactive and can be detected

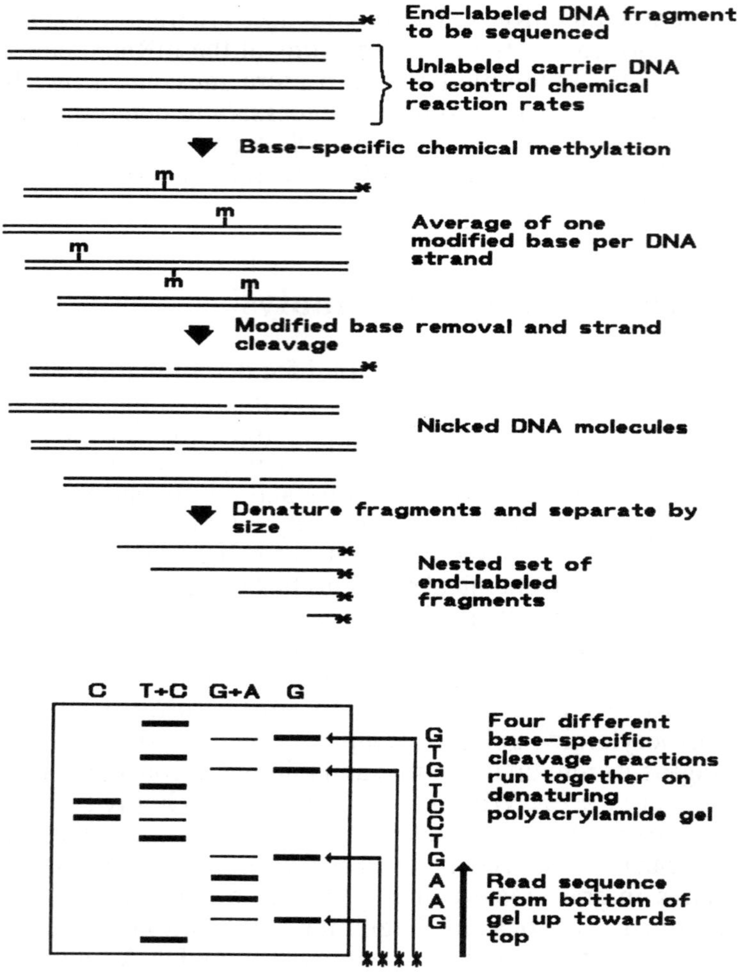

Figure 9.6. Chemical method of nucleotide sequence analysis. One end of the DNA fragment to be sequenced is labeled (usually with a ^{32}P-nucleotide) and mixed with an unlabeled carrier DNA used to control the chemical reaction rates of subsequent steps. DNA is then subjected to at least four different chemical reactions that modify (usually by methylating) the DNA in a residue-specific manner. Reactions used might modify at C residues, at both C and T residues, at both G and A residues, or at G residues alone. Chemical conditions are designed to modify an average of one residue per labeled DNA strand and are fairly random over the length of the DNA, so that each

by autoradiography. The 2',3'-dideoxynucleotide triphosphates (ddATP, ddCTP, ddGTP, ddTTP) are compounds that are similar in structure to the normal deoxynucleoside triphosphates but are missing the 3' hydroxyl group. These compounds can be incorporated into a growing DNA chain by DNA polymerase, but cause chain termination because the lack of the 3' hydroxyl group prevents formation of a 5'-3' phosphate bond with the next residue in the growing DNA chain. When a primed template is used in four different synthesis reactions, each in the presence of a different ddNTP, reactions that terminate specifically at A, C, G or T will be obtained. The DNA fragments present in these four reactions can be denatured, subjected to polyacrylamide gel electrophoresis and autoradiography, and the nucleotide sequence of the DNA fragment read directly off of the gel (Figure 9.4).

Although the label used for detection of the newly synthesized DNA has traditionally been either α ^{32}P-dATP or α ^{35}S-dATP, a variety of non-radioactive DNA detection systems have been developed to allow non-radioactive sequencing of DNA. While some of these non-radioactive methods utilize a staining procedure to detect the DNA sequence, the most powerful methods rely on fluorescent dyes coupled to either the sequencing primers or to the dideoxynucleoside triphosphates used in the chain termination reactions. Automated DNA sequence machines that are based on the detection of fluorescent analogs can

nucleotide of the sequence will be modified in at least one of the fragments in the reactions. The samples are then subjected to a second chemical reaction that removes the modified base and nicks the DNA strand at the site of base removal. The nicked DNA samples are then electrophoresed through a denaturing polyacrylamide gel that separates the fragments by size. Samples are visualized, generally by placing the gel against a sheet of X-ray film, to obtain a pattern of bands corresponding to the sequence of the nucleotide residues in the end-labeled DNA fragment. The sequence is read by starting at the bottom of the gel with the fastest-migrating band, corresponding to the smallest labeled DNA fragment, determining which chemical reaction generated that band, and designating the appropriate residue. G residues will form bands in the G+A and the G reactions, but A residues will only form bands in the G+A reaction.

2' deoxyribonucleoside triphosphate

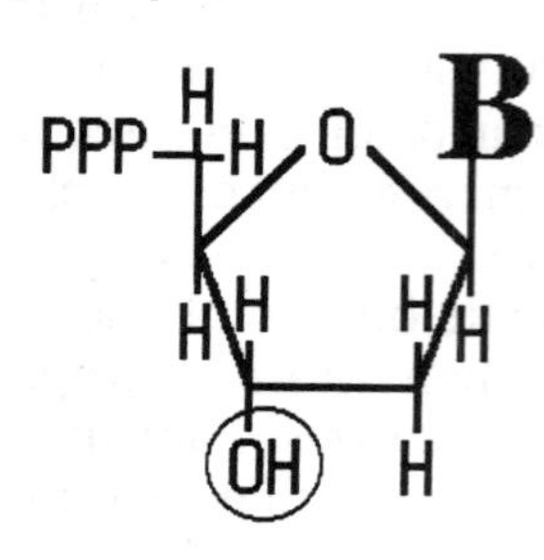

2',3' dideoxyribonucleoside triphosphate

PPP H H O B H H H H H H

a. ssDNA template

Oligonucleotide primer

b. ssDNA template

DNA polymerase

c. ssDNA template

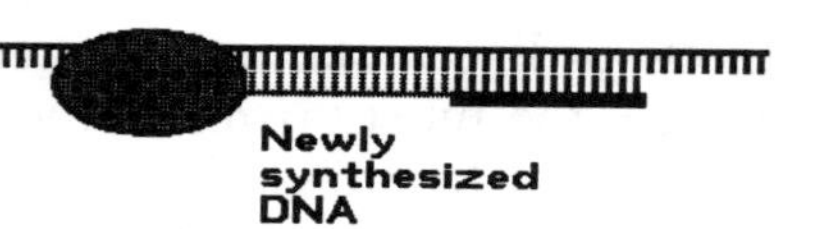

Newly synthesized DNA

d. ssDNA template

Terminated DNA fragment

Released DNA polymerase

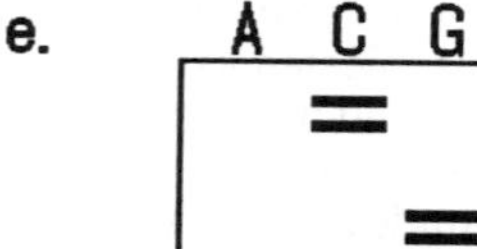

e.

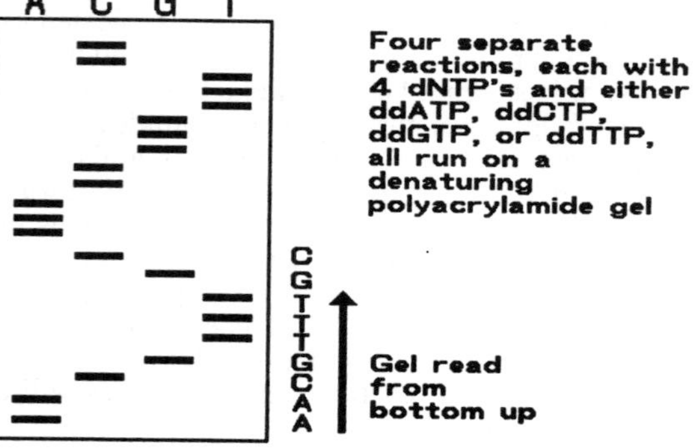

Four separate reactions, each with 4 dNTP's and either ddATP, ddCTP, ddGTP, or ddTTP, all run on a denaturing polyacrylamide gel

Gel read from bottom up

scan the sequence as the DNA fragments electrophorese through the polyacrylamide gel, interpret the sequence, and load the information directly into a computer. This innovation greatly reduces both labor and human error during nucleotide sequence analysis.

Sequencing large DNA fragments

With both sequencing methods, there is a limit to the amount of sequence that can be reliably determined with a single set of reactions. When the limit of sequence is 300 bases, it is obvious that reliable determination of the sequence of a fragment of 3000 bases would be difficult. This problem

Figure 9.7. The chain termination method of nucleotide sequence analysis is based on the ability of dideoxyribonucleoside triphosphates (ddNTP's) to be incorporated into a growing DNA chain and cause termination of elongation. The 2' deoxyribonucleoside triphosphates (dNTP's), the normal substrates for DNA synthesis, have a hydroxyl group at the 3' carbon of the sugar ring, allowing formation of an adjacent 5'-3' phosphate bond in a DNA chain. The ddNTP's have a hydrogen residue at the 3' carbon and cannot form an adjacent 5'-3' phosphate bond in a DNA chain, thereby causing chain termination. The sequence determination process involves:

(a) Annealing a synthetic oligonucleotide primer to the single-stranded template of interest
(b) Adding DNA polymerase (such as Klenow fragment or a modified bacteriophage polymerase), dNTP's with a label to allow detection of newly synthesized DNA, and a small amount of either ddATP, ddCTP, ddGTP, or ddTTP
(c) Synthesis of a new DNA strand
(d) Sequence-specific termination of the newly synthesized DNA chains
(e) The reactions are electrophoresed on a denaturing polyacrylamide gel.

A band that appears in the ddATP reaction was caused by incorporation of ddA and must correspond to the position of an A residue in the normal sequence. Similar logic applies to bands present in the ddCTP, ddGTP, and ddTTP reactions. The sequence of the DNA fragment can be directly read from the gel starting at the bottom with the smallest DNA fragment and proceeding upwards one band (corresponding to one base) at a time.

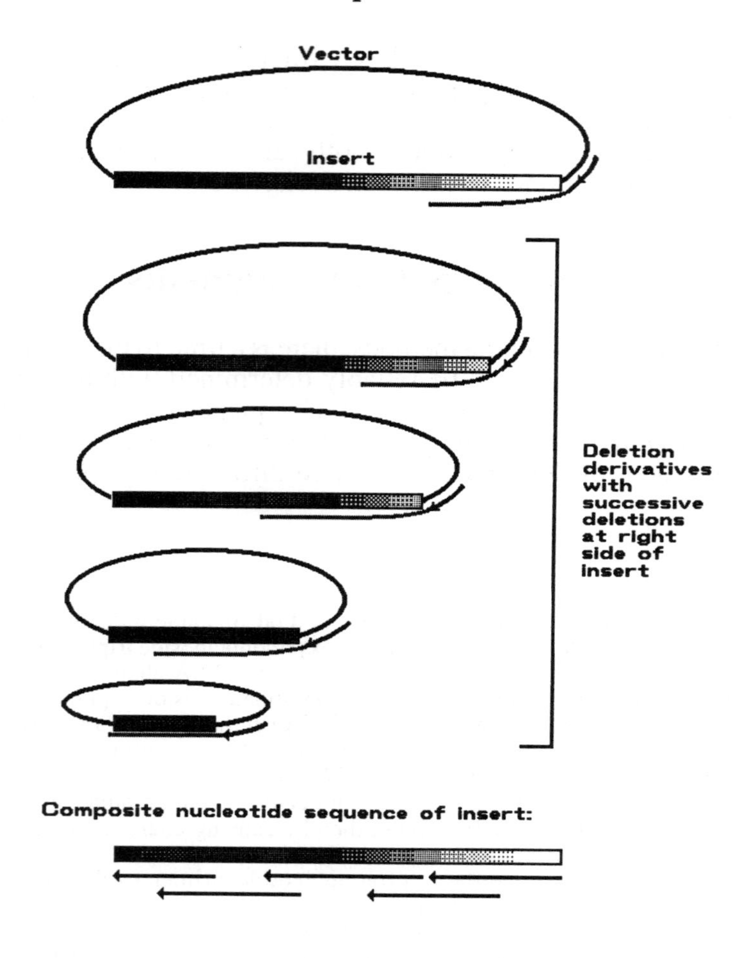

Figure 9.8. Because nucleotide sequence reactions are only reliable for a few hundred base pairs in length, the nucleotide sequence of a DNA fragment thousands of base pairs long cannot be determined with a single reaction. Determining the nucleotide sequence of a large DNA fragment can be accomplished by using enzymes to generate successive deletions at one side of the DNA fragment, isolating individual deletion derivatives, and sequencing each of the derivatives. Different amounts of the right-hand side of the insert in the figure have been removed to generate four smaller deletion derivatives. The sequence determined for the original insert overlaps with the sequence determined for the first deletion derivative, which then overlaps with the sequence determined for the second deletion derivative, and so on through the set of derivatives. A composite nucleotide sequence of the original insert can be made by aligning the five overlapping sequences.

can be circumvented by using protocols that successively delete regions that have already been sequenced and, in effect, walk the sequence start point through the fragment of interest (Figure 9.8). An enzyme such as exonuclease III can be used to digest away one end of a DNA fragment, followed by S1 nuclease and Klenow treatment to make the ends of the fragment blunt. The deletion derivatives are then cloned, isolated, and used as templates for sequencing reactions. For a 3000-base DNA fragment, 20 deletion derivatives each about 150 bases smaller than the nearest larger fragment might be sequenced to generate the nucleotide sequence of the entire fragment.

With the availability of DNA synthesis machines that can chemically synthesize oligonucleotides of a specific sequence, nucleotide sequence analysis can also be accomplished in a very methodical approach (Figure 9.9). Sanger sequence determination reactions can be performed on a cloned DNA fragment and the sequence determined as far as the reactions appear reliable. The last 20 to 25 bases that can be determined are then entered into a DNA synthesizer to make an oligonucleotide to be used as a new DNA sequencing primer. Sanger sequence determination reactions are then performed using the newly synthesized oligonucleotide as a primer for the Sanger synthesis reactions. The next portion of the sequence is determined, a third oligonucleotide primer synthesized, and the sequencing reactions repeated. This process can be continued until the nucleotide sequence of the entire fragment is completed.

Determination of gene structure

Although nucleotide sequence determination is a key element to understanding a gene, the nucleotide sequence alone is rarely sufficient to understand the structure of a gene. Many eukaryotic genes are not colinear but are punctuated with intervening sequences called introns. The nucleotide sequences that actually code for a protein, termed exons, may be broken up into several blocks of sequence information, each of which may be separated from the other

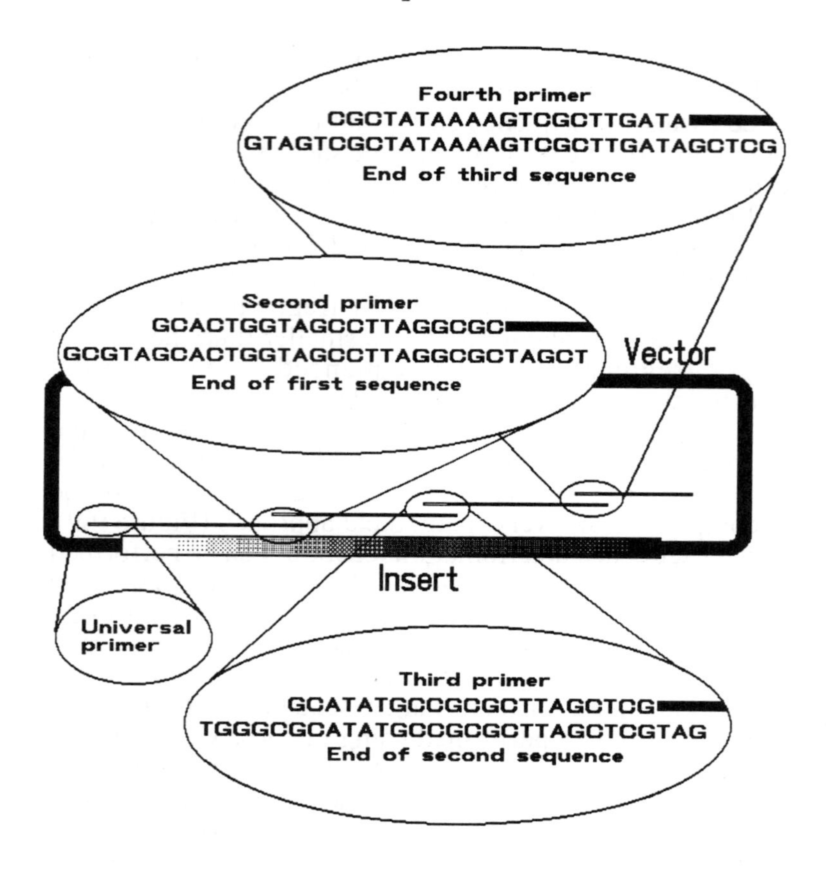

Figure 9.9. Synthesis of new primers as a sequencing strategy. Large DNA fragments can be sequenced by synthesizing DNA sequencing primers that initiate DNA sequencing reactions at new points successively farther into the DNA fragment. Sequencing may begin with a "Universal" primer that anneals adjacent to the cloning site in the vector and directs DNA sequencing reactions into the DNA insert. The first sequence obtained is read as far as possible and the end of the sequence is used to design a second sequencing primer. The second primer is then used to direct sequencing reaction from the spot where it anneals farther into the DNA insert. The end of the second sequence is then used to synthesis a third primer for another DNA sequencing reaction. The process of prime, sequence, read, synthesize primer is repeated until the entire sequence of the insert is complete and the sequence reactions continue into the vector DNA. The composite of all of the sequence reactions together yields the nucleotide sequence of one strand of the DNA insert under investigation.

```
cDNA sequence:

        ATGAGCTCGCTGCTTTAGGATGCATATTTTTATATAAAAGGACGCTCGTCGAAAAAAAAAAA

Genomic DNA sequence:

    ATAGGGGATAATGAGCTCGCTGCTTTAGAGATTTTAGATTATATCGCTCGATGCATATTTTTATATAAAAGTAAGCGCGCATAAGACGCTCGTCGTAGCATGCTTGTT

Aligned sequences:
                                                                          PolyA tail
                                                                    AAAAAAAAAAAAA
    ATGAGCTCGCTGCTTTAG                          GATGCATATTTTTATATAAAAGGACGCTCGTCG
ATAGGGGATAATGAGCTCGCTGCTTTAGAGATTTTAGATTATATCGCTCGATGCATATTTTTATATAAAAGTAAGCGCGCATAAGACGCTCGTCGTAGCATGCTTGTT

        Exon 1       Intron 1                 Exon 2
```

Figure 9.10. Alignment of the cDNA and genomic sequences can identify the protein coding regions called exons and the intervening sequences, or introns, that are initially transcribed but removed prior to translation.

exons by thousands of bases of an intron. The intron/exon structure of a genomic sequence must be determined as an important aspect of understanding gene structure.

The mRNA molecule, rather than the DNA template from which the RNA molecule was synthesized, is actually used by ribosomes as the template for the assembly of amino acids into proteins. The sequence of an mRNA molecule consists of the exons assembled into a linear sequence with the introns removed. The sequence of an mRNA molecule is generally determined by cloning the cDNA corresponding to that mRNA, then determining the nucleotide sequence of the cDNA. Since the cDNA sequence consists of the exons alone, direct comparison of the cDNA sequence with the sequence of the genomic clone will reveal the location and the sequence of the genomic introns (Figure 9.10).

S1 nuclease mapping

Since a eukaryotic gene can contain many introns and be over 100,000 base pairs long, sequencing the entire genomic sequence can be a tedious approach to determining intron/exon pattern. A technique called S1 nuclease mapping has also been used to examine the number of introns in a genomic clone (Figure 9.11). The genomic fragment (gDNA) is allowed to hybridize with mRNA or cDNA corresponding to the gene of interest. Following annealing, the gDNA:mRNA or gDNA:cDNA hybrid is treated with S1 nuclease, which digests the single-stranded genomic intronic regions that have not annealed to mRNA or cDNA. Where an intron is present in the genomic sequence but not present in the mRNA or cDNA, S1 treatment will remove the intron sequences from the genomic DNA. The treated DNA can be denatured and examined by gel electrophoresis, where fragments can be sized. Each exon will be detected as a specific DNA fragment. The number of specific fragments will identify the number of exons and therefore the number of introns in the gene.

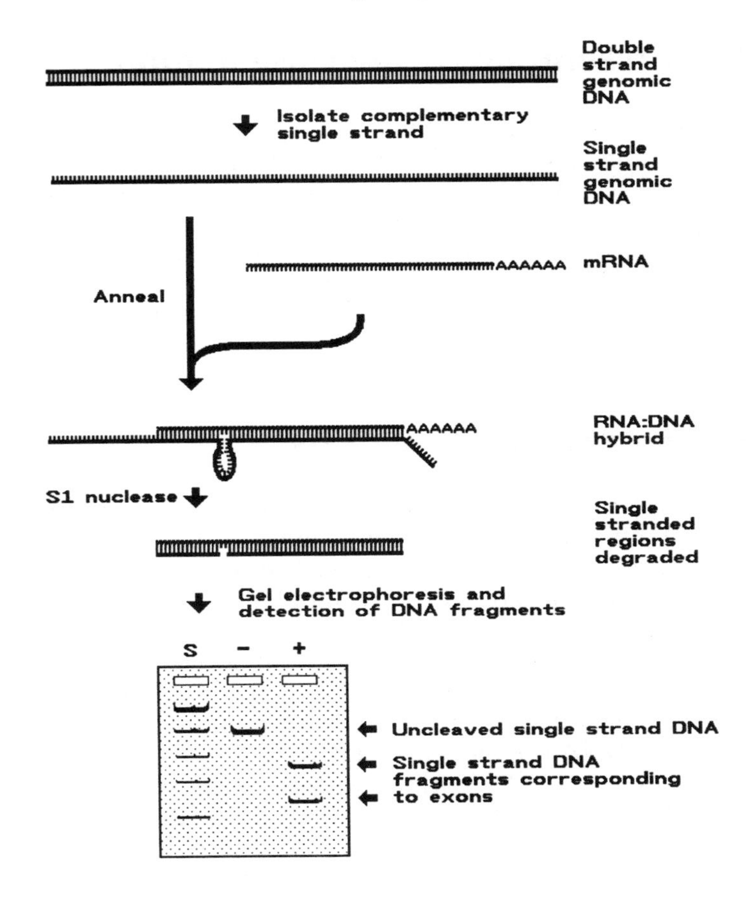

Figure 9.11. S1 nuclease analysis can be used to identify regions of mismatch between a DNA fragment and an RNA molecule synthesized from that DNA fragment in the cell. A cloned genomic DNA is used to make a single-stranded DNA complementary to the corresponding mRNA molecule. The DNA and mRNA are allowed to anneal to form a DNA:RNA hybrid. Regions of DNA that are not present in the mRNA molecule form single-stranded tails or loops in the DNA:RNA hybrid. The hybrid is treated with S1 nuclease, which degrades only the regions of single-stranded nucleic acid, removing the tails and loops and leaving a break in the DNA at the site of the unpaired loop. Following gel electrophoresis and detection of the DNA fragments, the S1 nuclease-resistant DNA fragments correspond to the regions of DNA that are present in the mature mRNA molecule. This approach can be used to identify number and approximate location of introns and exons in the absence of any nucleotide sequence information.

High resolution S1 nuclease mapping

The start point or the terminus of an mRNA molecule can be determined by a modification of S1 nuclease mapping (Figure 9.12). A DNA fragment that overlaps the end of the mRNA is allowed to anneal with mRNA, then digested with S1 nuclease. The protected double-stranded DNA:RNA hybrid can be sized by polyacrylamide gel electrophoresis, or chemically sequenced to determine the start point of the mRNA molecule.

Primer extension analysis

The start point of an RNA molecule can also be determined by a technique called primer extension, similar to the process used to synthesize cDNA from mRNA (Figure 9.13). A small, radioactively labeled DNA primer is annealed within the mRNA molecule and reverse transcriptase is used to elongate the primer. The elongation will proceed to the end of the mRNA template. The resulting DNA is denatured, sized by gel electrophoresis, and the elongated DNA fragment detected by autoradiography. As the location of the DNA primer within the mRNA molecule is known from sequence analysis, the length of the elongated primer can be used to determine the end of the mRNA molecule and hence the transcription start site.

Understanding regulation of gene expression

One of the major goals of molecular biology is to understand the processes that control the expression of genes during the development of an organism. This problem is particularly intriguing because the cells in different tissues of a complex organism contain essentially the identical chromosomal genes, yet cells in different tissues express different subsets of the total available genes.

Nucleotide sequence analysis can reveal the protein

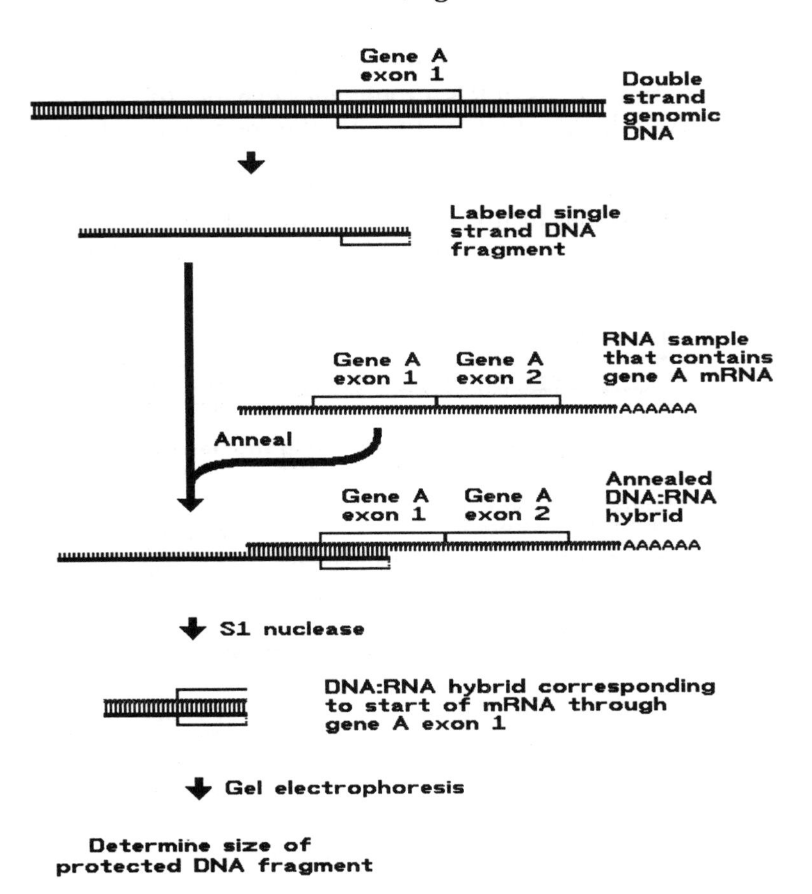

Figure 9.12. High resolution S1 nuclease analysis used to identify the 5' terminus of an mRNA molecule. A labeled single-strand DNA fragment corresponding to part of exon 1 and the 5' non-coding region of a gene (gene A) is annealed with an mRNA sample that contains the mature gene A mRNA. The annealed mRNA:DNA hybrid consists of a double-stranded region with a single-stranded tail at each end. The hybrid is treated with S1 nuclease to destroy the single-stranded regions and then electrophoresed on a denaturing polyacrylamide gel with molecular size standards to determine the length of DNA protected in the hybrid. The size of the protected fragment indicates the distance from the DNA end located within gene A exon 1 to the 5' terminus of the mRNA molecule.

coding information stored in DNA molecules and analysis of the 5' and 3' ends of RNA molecules can identify regions of DNA molecules that are important to regulation of gene expression. These methods alone, however, are not sufficient to allow examination of the molecular mechanisms that turn genes on and off. Although nucleic acids store the genetic information of a cell, interactions of

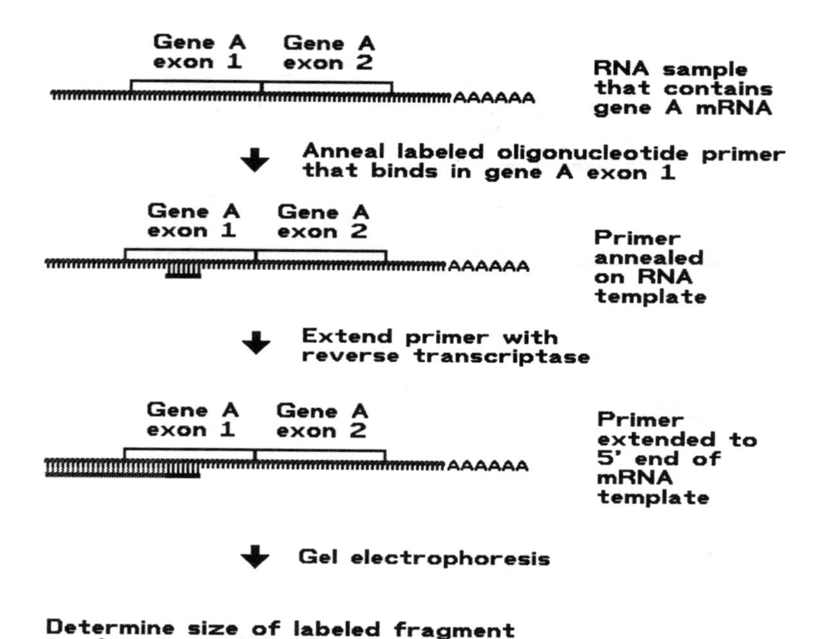

Figure 9.13. Primer extension to determine a 5' mRNA terminus. A labeled single-stranded oligonucleotide primer that has been either chemically synthesized or isolated from a larger cloned DNA fragment is annealed with an RNA sample that contains the mature gene A mRNA. The enzyme reverse transcriptase and all four dNTP's are then added. The enzyme uses the primer as the starting point for DNA synthesis and the mRNA strand as a template and synthesizes DNA to the end of the mRNA molecule. The reaction is electrophoresed on a denaturing polyacrylamide gel with size standards and the extended labeled primer is visualized. The size of the extended primer indicates the distance from the annealing site of the primer to the 5' terminus of the mRNA molecule.

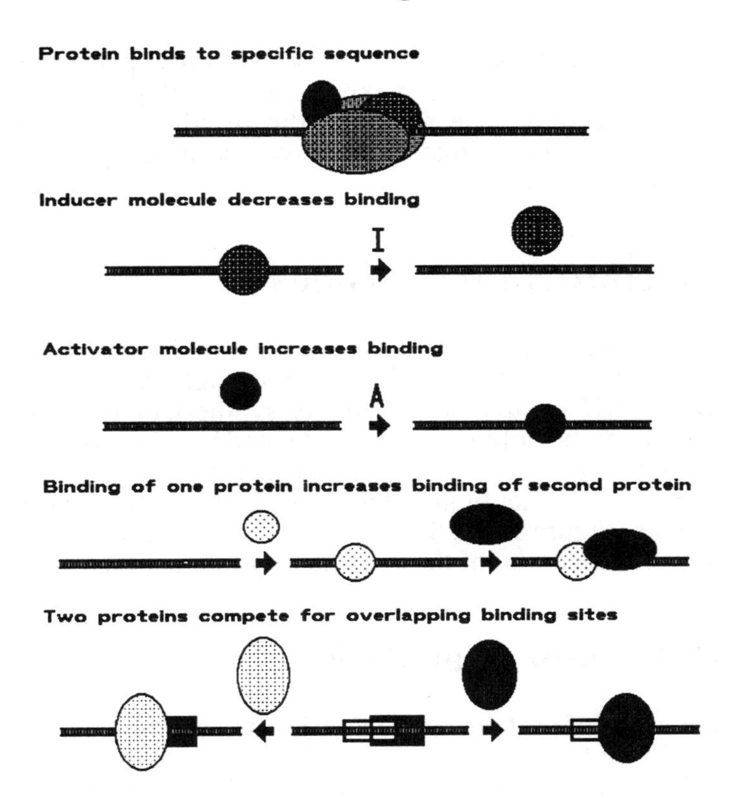

Figure 9.14. Study of the regulation of gene expression involves understanding how DNA binding proteins interact with both DNA and other regulatory factors. DNA binding proteins can recognize and bind to DNA in a variety of manners. A relatively simple DNA binding protein, such as a restriction endonuclease, may recognize and bind at a specific sequence. Many of the proteins involved in regulating efficiency or specificity of transcription have more complicated interactions. The prokaryotic negative regulatory proteins called repressors bind to specific sequences but can also bind small molecule co-factor called inducers that reduce binding affinity and cause release of the protein from the DNA. Positive regulatory factors, proteins that stimulate initiation of transcription, often bind tightly to DNA only after binding a small molecule activator. Regulation of gene expression can involve the interactions of many different proteins with the same general region of DNA. The binding of one protein may facilitate or increase the binding of a second protein, and two proteins may compete for binding at different sequences that overlap one another. Regulation of gene expression is ultimately the result of interaction of a number of different DNA binding proteins with DNA, with each other, and with small molecules present in the cell.

proteins with nucleic acid and other proteins appear to be the key to regulating gene expression in both prokaryotes and eukaryotes. The study of the regulation of gene expression must ultimately focus on the investigation of protein:nucleic acid interactions.

Protein:DNA interactions

Proteins modulate gene expression by binding to specific sequences in a DNA molecule. Although the complex enzyme RNA polymerase actually synthesizes RNA, the function of the polymerase is modulated by a host of DNA binding proteins, including both positively and negatively acting factors, that bind to DNA or to other DNA binding proteins (Figure 9.14). Several methods have developed to allow visualization of these regulatory interactions.

DNA footprint analysis

Interaction of proteins with DNA is often examined by the technique of footprinting (Figure 9.15), in which a protein is allowed to interact with a ^{32}P end-labeled DNA fragment and modify the access to regions of the DNA helix. The protein-DNA complex is then partially fragmented by treatment with DNAse I or by chemical sequence reactions and the resulting fragments examined by gel electrophoresis. Interaction of protein with the DNA helix can alter the accessibility of the DNA and thereby alter cleavage during the chemical or enzymatic reactions: some bases may be protected from cleavage, while cleavage at others may be enhanced. The region where cleavage of DNA is altered by the binding of the protein is called the "footprint" of the protein and identifies specific DNA sequences where protein interaction may be important to regulation of gene expression.

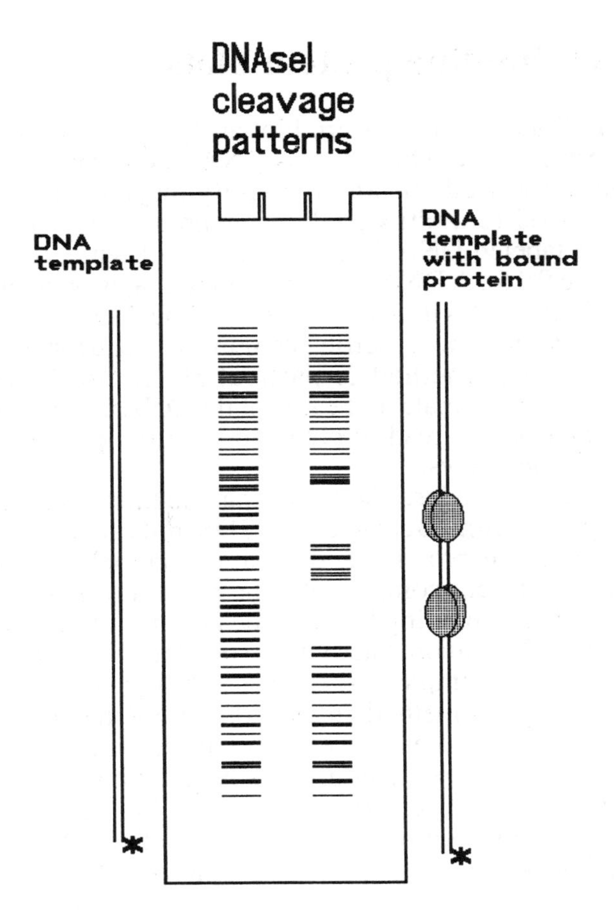

Figure 9.15. The interaction of proteins with DNA can be directly visualized with a technique called DNA footprinting. An end-labeled DNA template that is either treated with the endonuclease DNAse I or treated with chemical sequence reactions can be partially cleaved to generate a ladder of end-labeled fragments on a denaturing polyacrylamide gel, as illustrated at the left of the figure. If a DNA binding protein is allowed to bind to the DNA template prior to the cleavage reactions, the regions of DNA that are bound by the protein will have altered suseptibility to the cleavage reactions. As illustrated at the right of the figure, regions that are completely covered by the protein may not cleave at all and other regions may actually be cleaved more readily than in the absence of the DNA binding protein. This region of altered cleavage is called the footprint of the protein on the DNA template and allows direct visualization of protein:DNA interaction.

DNA-binding protein blots

A modification of Western blot analysis, often referred to as a Southwestern blot, allows DNA-binding proteins to be first separated on an SDS-polyacrylamide gel and transferred to a nitrocellulose membrane, then incubated with a labeled DNA fragment instead of with an antibody (Figure 9.16). A DNA binding protein that is present on the membrane and is in an active state will bind a DNA fragment that contains the specific recognition sequences for the protein. The method is particularly useful because it provides an estimate of the size of the DNA binding protein, information not readily obtained with footprinting or gel retardation assays.

This procedure, however, has distinct limitations in its application. Because the SDS-polyacrylamide gel denatures the proteins and separates complexes into the separate protein components, the protein of interest must be capable of binding DNA in the absence of other protein factors. DNA binding activity associated with proteins that are part of a complex, like RNA polymerase, may not be detected by this method. In addition, SDS denaturation prior to gel electrophoresis is a relatively harsh treatment and may irreversibly denature many proteins. Unlike the antibody detection procedures in a Western blot, the Southwestern requires the presence of active DNA-binding proteins on the nitrocellulose membrane. Consequently, proteins that cannot renature to an active form will also remain undetected by Southwestern blot analysis. Many well-characterized DNA-binding proteins cannot be detected by Southwestern analysis.

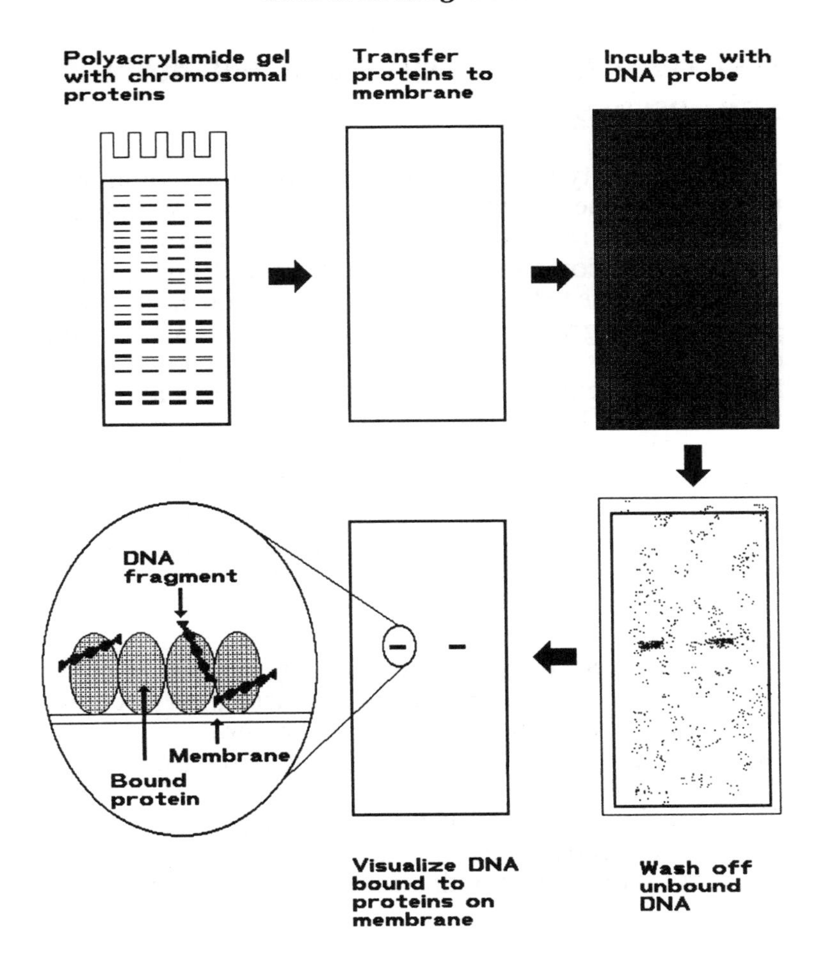

Figure 9.16. DNA-binding protein blots. Southwestern or DNA-binding protein blot analysis uses the same basic approach as a Western blot to separate chromosomal proteins on an SDS-polyacrylamide gel, then transfer the proteins to a membrane. The blot is incubated with a labeled DNA probe (usually double-stranded) that is believed to be specifically bound by a DNA binding protein. During the incubation period, the proteins on the membrane may renature or fold up into their active shapes and regain DNA binding activity. When the excess unbound DNA is washed off the blot, the bands corresponding to proteins that bind the DNA probe can be visualized. While this method does not work with all DNA binding proteins, it facilitates investigation of proteins that are functional with this assay.

Gel retardation assays and gel mobility shift assays

The gel mobility shift assay is a method more commonly used to examine the interaction of DNA-binding proteins. This procedure is based on the observation that a DNA:protein complex has an altered gel electrophoretic mobility relative to that of the DNA alone, with the complex generally migrating much more slowly. A purified oligonucleotide or small DNA fragment is generally end-labeled and incubated with a protein extract to allow the formation of DNA:protein complexes. Samples are examined by gel electrophoresis and the retarded complexes visualized by detection of the labeled DNA (Figure 9.17). Incubation conditions must be carefully controlled and resistance of complexes to salt, specific DNA competitors, and nonspecific DNA competitors must be examined to understand the significance of various DNA:protein complexes.

These reactions are much more versatile than Southwestern blot analysis because the incubation reactions involving DNA and protein can be more carefully manipulated to maintain proteins in an active conformation. In addition, these assays can reveal the presence of multiple DNA:protein complexes that can occur when a DNA has more than one protein binding site or when multiple proteins compete for or interact with a common binding site.

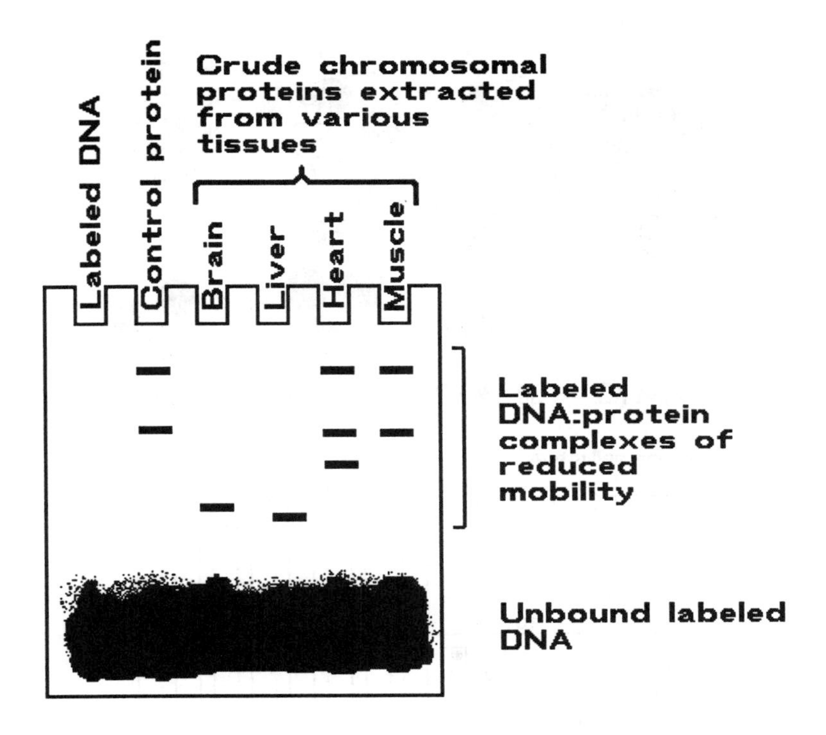

Figure 9.17. Gel mobility shift assays. This assays use the principle that an oligonucleotide that contains the binding site for a protein can be labeled, then incubated with an extract containing the DNA-binding protein to allow the formation of a DNA:protein complex. When the reactions are subjected to electrophresis of a polyacrylamide gel, the DNA:protein complex has both altered charge and net mass that retard mobility relative to that of the unbound DNA. Visualization of the location of the labeled DNA will reveal large amounts of unbound DNA and discrete bands corresponding to the DNA:protein complexes. In the example illustrated, a DNA fragment has been incubated with a purified control protein known to bind and with crude chromosomal proteins extracted from brain, liver, heart, and muscle. The results indicate that heart and muscle generated shifted complexes of the same mobility as the control protein complexes, while brain, liver, and heart generated shifted complexes of different mobilities.

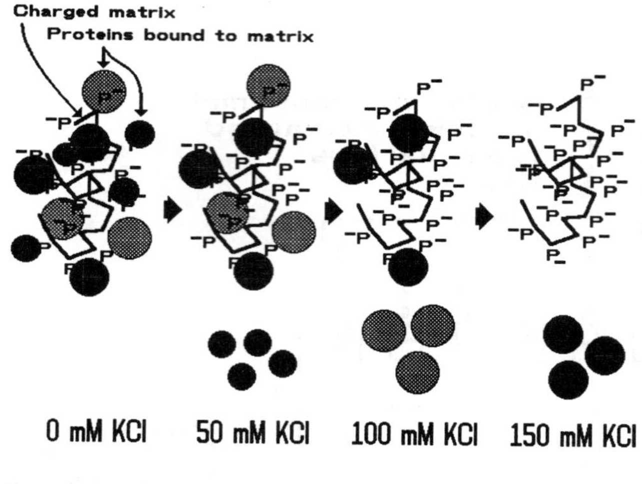

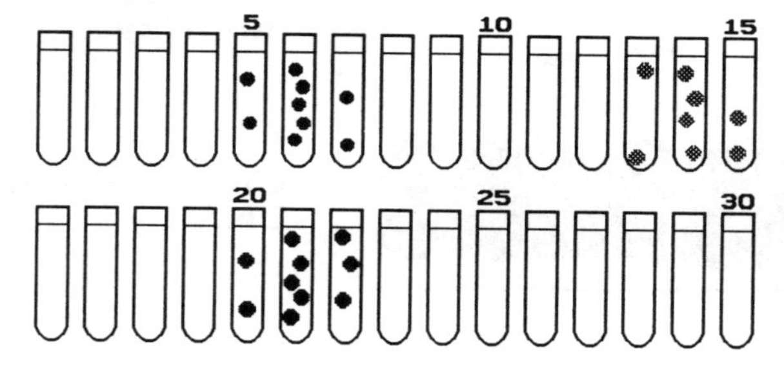

Figure 9.18. Purification of DNA-binding proteins by ion-exchange column chromatography. A charged support matrix like phosphocellulose, a cellulose polymer with phosphate residues (P^-) attached, is mixed under conditions of low ionic strength (low salt concentration) with a nuclear extract containing many different DNA binding proteins. Many of the proteins will interact with and bind to the charges on the matrix. As salt concentration is gradually increased, the charged salt molecules displace the charge interactions that keep proteins bound to the matrix. Each different protein will be released from the matrix at a salt concentration characteristic for that protein. If a gradual gradient of increasing salt concentration is passed across the matrix and fractions collected, each successive fraction will have a slightly higher salt concentration than the previous fraction. Since different proteins release from the matrix at different salt concentrations, different proteins will be present in different tubes. Ion-exchange and affinity column chromatography allow reasonably rapid purification of a desired protein away from other contaminants.

Purification of DNA binding proteins

One approach to understanding how proteins interact with DNA to regulate gene expression involves dissecting transcription complexes to identify and characterize the individual protein components. Gel retardation and Southwestern blots are frequently used as assays for monitoring DNA-binding activity during protein purification. Purification of DNA-binding proteins is generally accomplished by a combination of DNA affinity chromatography, ion-exchange chromatography, and gel filtration.

A DNA fragment that contains the binding site for a desired protein can be coupled to a support matrix to form a resin that will bind DNA-binding proteins (Figure 9.18). Under conditions of low ionic strength (low salt concentrations), many DNA-binding proteins will interact with this matrix. As salt concentration is increased, the weak protein:DNA interactions are destabilized and non-specific interactions are disrupted. Under conditions of high ionic strength, a DNA-affinity matrix will selectively bind only proteins that bind strongly to the DNA fragment coupled to the matrix. The affinity resin can therefore be used to purify a desired protein away from other DNA-binding proteins that do not react with the specific DNA fragment.

Ion-exchange chromatography allows a mixture of proteins to bind to a charged matrix, such as DNA cellulose, phosphocellulose, diethylaminoethylcellulose (DEAE), or carboxymethylcellulose (CMC). Under conditions of low ionic strength, the charges of proteins can interact with the charged matrix, causing the proteins to bind. As ionic strength is increased, the charge interactions are destabilized, allowing proteins to be released from the matrix. Proteins that bind weakly elute from the matrix at a lower salt concentration than proteins that bind strongly, allowing separation of proteins with different charge properties. A mixture of proteins can be bound to a column of the charged support matrix and eluted with a salt gradient, slowly increasing the salt concentration on the column. Fractions collected as the salt gradient elutes from the column will contain different proteins.

Both DNA-affinity and ion-exchange chromatography operate on the principle of using increasing salt to destabilize interaction of proteins with a matrix. Gel filtration (Figure 9.19) separates proteins based on their size. The filtration matrix consists of beads containing pores of a specific size. When these beads are poured into a column, proteins applied to the top of the column and pushed through the column with a a flow of buffer have two possible routes to reach the bottom of the column. Proteins that are too large to enter the pores will be excluded from the matrix and will pass around the beads, while the smaller proteins will enter the pores and migrate through the beads. Proteins ultimately move through the column according to their size, with large proteins emerging from the column prior to the smaller proteins.

Typical DNA-binding protein purification schemes involve the use of all three of these protein purification methods to take advantage of the different physical properties of different proteins.

In vitro transcription assays

One approach to understanding regulation of gene expression involves reproducing or reconstituting the RNA synthesis complex in a test tube. Purified RNA polymerase and the four ribonucleoside triphosphates can be combined with a DNA template in a tube and any resulting RNA examined to determine whether the *in vitro* transcription product mimics the natural RNA molecules synthesized *in vivo*. While transcription from many prokaryotic promoters can be readily duplicated in this manner, the requirements are more strict for transcription from most eukaryotic promoters. *In vitro* transcription reactions often require the addition of several purified transcription factors for duplication of the features of *in vivo* regulation of gene expression. These reconstituted transcription systems are an important tool in elucidating the biochemical events involved in gene expression.

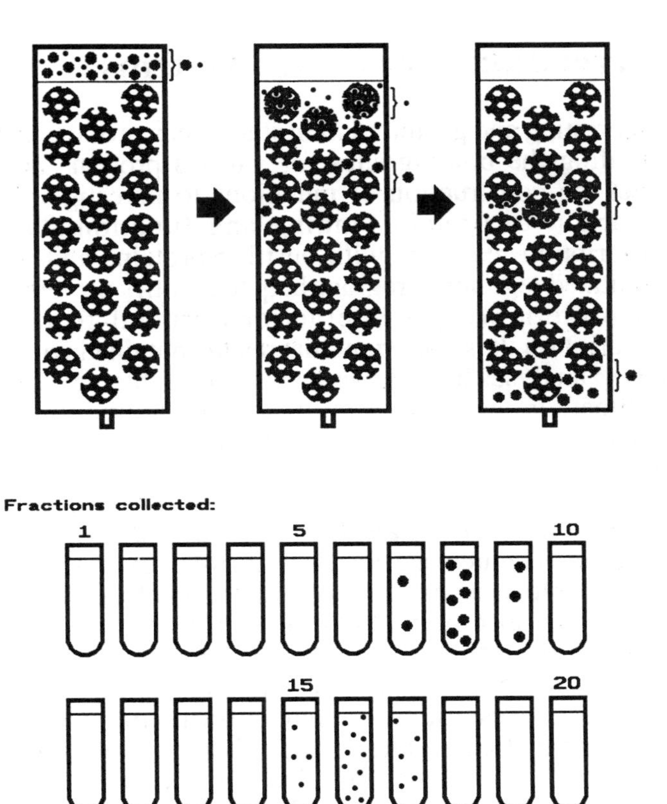

Figure 9.19. Gel filtration as a method of separating large molecules according to size. A sample containing proteins of different sizes can be applied to a column of tiny porous beads that has been poured in a glass or plastic cylinder. As buffer is applied to the top and collected from the bottom of the column, the proteins will be carried along through the beads by the buffer flow. Molecules that are too large to enter the pores in the beads will travel around the beads and rapidly move through the column. The smaller molecules will enter the pores in the beads and must move through the beads, rather than around the beads, on the way to the bottom of the column. The smaller proteins consequently move more slowly than the larger proteins. As fractions are collected at the bottom of the column, proteins will elute in decreasing order of size, with largest first and smallest last. Gel filtration can be used to purify proteins according to size or to remove small molecules like salts from large molecules like proteins and DNA.

Correlation with *in vivo* systems

Footprinting and gel retardation studies can provide insights regarding the binding sites for various proteins and can direct the construction of mutations to investigate bases important to protein binding and function. *In vitro* transcription reactions with purified proteins and normal or mutated regulatory regions help to identify protein:DNA and protein:protein interactions crucial to *in vitro* transcription. It is important that results obtained with these *in vitro* analyses be compared with those obtained with the legitimate *in vivo* transcription complex found in the cell to be reasonably certain that the events observed in the test tube are a good representation of cellular processes.

Many of the methods used to investigate gene expression *in vivo* have already been discussed. For example, S1 nuclease analysis, primer extensions, and Northern blots can give a good indication of transcript start point, tissue specificity, and level of transcription. Many of the most informative *in vitro* observations involve the investigation of the effects of mutations on the regulation of gene expression. It is particularly important that mutations that cause alterations in function of an *in vitro* transcription complex be introduced into a cell to determine whether these mutations cause alterations in gene expression *in vivo*.

Fusion of regulatory regions to reporter genes

It is relatively easy to mutate the regulatory region of an isolated gene and introduce the derivative into the appropriate host to investigate the effect of the mutation on regulation of gene expression. When the mutated gene is introduced back into a cell that already contains a normal chromosomal copy of the gene, it is important that expression of the mutated gene can be distiguished from expression of the normal chromosomal copy. Two general strategies can be used to ensure detectability of the introduced gene: the mutated gene can be introduced into a host deficient in the chromosomal copy of the gene or the regulatory region of the gene under investigation can be attached to a reporter gene, a gene not normally expressed in the cell of interest.

Although deletion of the chromosomal copy of a gene can often easily be accomplished in prokaryotes and in some simple eukaryotes, like yeast, deletion cannot always be accomplished without affecting cellular physiology. Alterations in physiology may in turn affect expression of the introduced gene, leading to incorrect experimental conclusions. Although feasible, construction of deletions in higher eukaryotes can be slow and too expensive for routine investigations.

Fusion of a regulatory region of interest to a reporter gene allows regulation to be examined without the need to inactivate expression of the normal chromosomal gene (Figure 9.20). A reporter gene can be any gene that has an easily assayed gene product that is not generally expressed in the cell of interest. Regulatory regions are often fused to the gene encoding chloramphenicol acetyltransferase (*cat*). Gene expression can be evaluated by measuring the production of chloramphenicol acetyltransferase, a protein that can acetylate chloramphenicol in a simple enzymatic assay. Regulatory genes can be mutated, attached to reporter genes, and transformed into host cells to examine the *in vivo* effects of the mutations of gene expression. Correlation of the *in vivo* observations with results obtained with purified proteins can help validate the *in vitro* results.

Construction of chromosomal integrates and transgenic animals

Transformation of cells with regulatory regions fused to reporter genes does not always give an accurate representation of the regulatory interactions that take place when the same regulatory region is present in chromosomal DNA. The expression of a gene can be profoundly influenced by distal sequences as much as several thousand

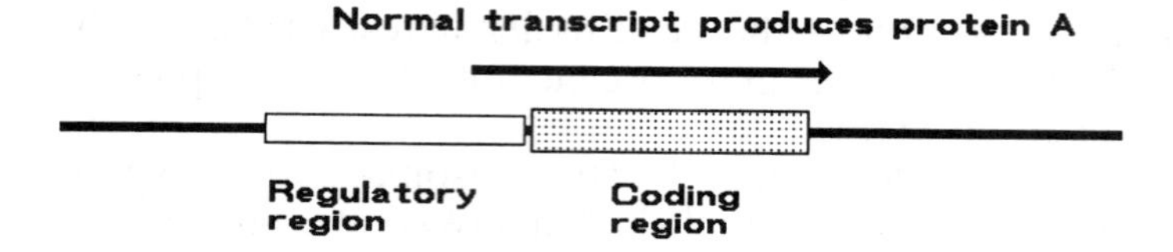

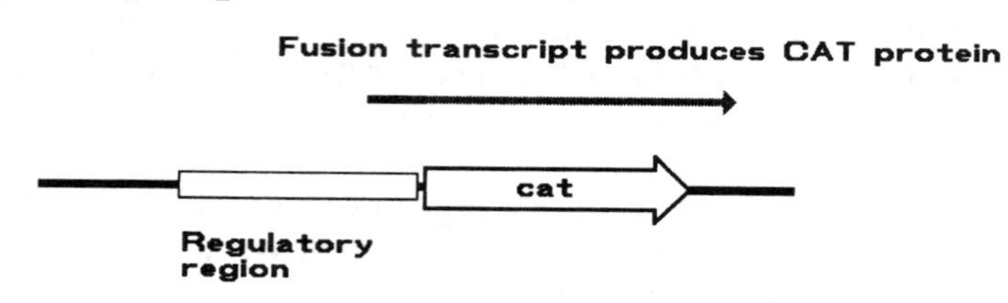

Figure 9.20. Fusions of regulatory regions to an easily assayed reporter gene allow convenient investigation of the role of DNA sequences in initiation of transcription. Restriction enzymes can be used to fuse the regulatory region of a gene of interest to a promoterless reporter gene, such as the *cat* gene. When the DNA construct is transformed into a host cell, transcription initiating at the cloned regulatory region will cause the synthesis of a fusion transcript that codes for the production of CAT protein. Since CAT activity is not normally present in most cells, the amount of transcription initiating at the cloned regulatory region can be estimated by determining amount of CAT activity produced by cells containing the regulatory region:reporter gene construct.

base pairs away from the start of transcription. When these distal sequences are not present on the isolated regulatory region fused to the reporter gene, the regulatory properties of the DNA construct will be different than those of the actual gene.

It may be important to integrate the regulatory region:reporter gene fusion into the chromosome. Since some regions of chromosomes are expressed at higher levels than other regions, insertion of a DNA construct into different sites or chromosomal positions can lead to different regulatory features. Positional effects are well-documented, particularly in plants, where insertion of a gene into different chromosomal locations can result in ten-fold differences in levels of gene expression. One of the most technologically impressive developments in the reintroduction of genes back into host organisms has been in the construction of transgenic animals, where the nucleus of a eukaryotic cell first functions as the recipient for a gene construct, then is inserted into an enucleated egg to allow development of a complete animal. Transgenic animals can allow investigation of the function of an altered gene in the appropriate context of cellular physiology.

Role of molecular techniques in characterizing genes

The methods that allow the isolation, amplification, and identification of DNA fragments are the basic tools used by molecular biologists in the attempt to understand the structure and function of genes. Although these methods are often remarkably simple in concept, the application of several individual methods in concert or in series allows very detailed genetic investigation. These molecular tools are extremely powerful and their application has resulted in a veritable explosion of knowledge concerning the structure and function of living cells.

Summary

1. **A primary goal of genetic research is the understanding of gene structure and gene regulatory mechanisms.**

2. **Nucleotide sequencing is used to determine the chemical structure of an isolated gene and the corresponding mRNA and helps reveal genetic information stored in DNA.**

3. **Comparison of mRNA structure with genomic DNA structure helps clarify gene structure by providing information about introns and exons.**

4. **Analysis of protein-DNA interaction is fundamental to the study of gene regulation. The interaction of proteins with DNA can be directly visualized by several types of *in vitro* assay.**

5. **It is important to compare *in vitro* observations with *in vivo* patterns of gene expression to be certain that results obtained with purified components are indicative of regulatory interactions in the cell.**

10

PCR Manipulation of Nucleic Acids

"Go away, boy, you bother me!" — W.C. Fields

Although recombinant DNA technology proven extremely powerful in the isolation and analysis of the structure and function of nucleic acids, many of the methods came to be considered slow and tedious. Although automation was able to improve some of the more monotonous tasks, such as preparation of miniscreen DNA and nucleotide sequence analysis, a great deal of recombinant DNA methodology was supplanted by the discovery of a process that allowed the amplification of a specific DNA fragment outside of any biological host. The speed and utility of this process, called the Polymerase Chain Reaction, has created a second technological revolution in the analysis of nucleic acids.

The Polymerase Chain Reaction (PCR)

The PCR process is based on the discovery that the DNA polymerases isolated from thermophiles, the organisms that live in hot springs, are not only active at very high temperatures, but can be subjected to repeated exposures to 100°C without losing activity. The heat-stable DNA polymerase *Taq*I purified from the bacterium *Thermus aquaticus* can be mixed with a DNA template, a synthetic

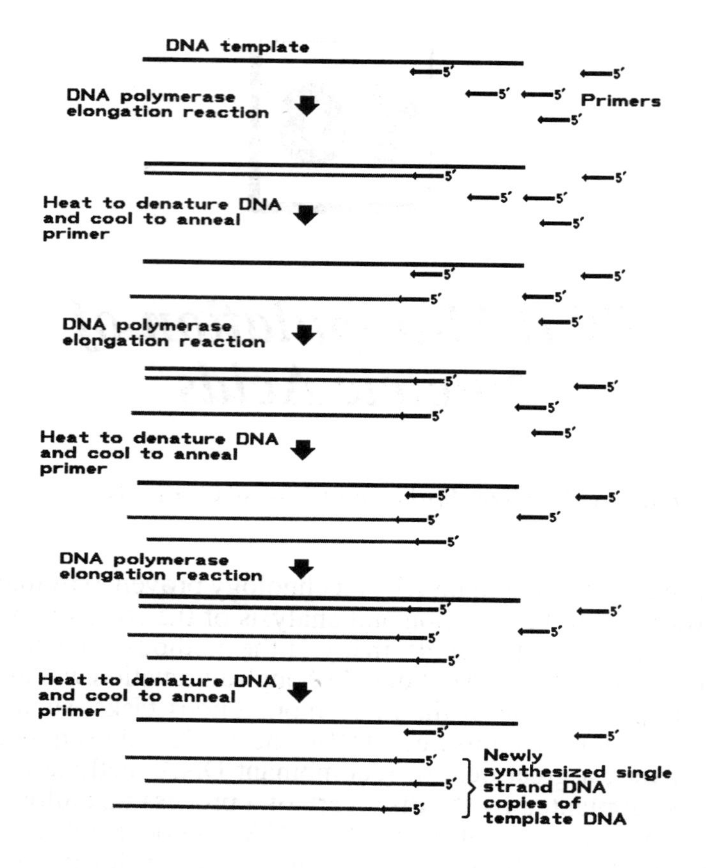

Figure 10.1. Cycling of thermostable polymerase. A reaction mixture containing a DNA template, an oligonucleotide primer that can anneal at a specific location on the DNA template, deoxynucleoside triphosphates, and a thermostable DNA polymerase can be heated to denature the DNA, then cooled to allow annealing of the primers. The polymerase can extend the annealed primer to the end of the template DNA, synthesizing a double-stranded molecule. Since additional DNA synthesis requires additional primed DNA template, the reaction cycle is then repeated. The sample is heated to denature the DNA and cooled to allow primer annealing. DNA synthesis again extends the primer to make a double-stranded molecule. Since the primer is complementary to the original DNA template but is the same sequence as the newly synthesized DNA, the newly synthesized DNA cannot be primed and remains single-stranded. As the cycle is repeated, single-stranded DNA molecules complementary to the original DNA template will accumulate with linear kinetics. The reaction can be repeated for up to 20 cycles without inactivating the polymerase.

DNA oligonucletide primer that anneals to the template, and the four deoxynucleoside triphosphates to make a reaction mixture. When the reaction is heated to 100°C for a brief period, the DNA will denature. When allowed to cool, the oligonucleotide primer will anneal to the complementary sequence in the DNA to form a primed template that can be elongated by the *Taq* DNA polymerase. This reaction is analogous to the elongation of a primer by Klenow fragment of DNA polymerase I during the chain termination method of DNA sequence determination. After a short synthesis period, a new copy of DNA results from elongation of the annealed primer. Because of the thermal stability of the *Taq* polymerase, this process can be repeated for many cycles with only a slight loss of enzyme activity during each cycle (Figure 10.1). Each cycle will result in priming of the DNA template and the subsequent synthesis of a new single-strand DNA copy. In the presence of a primer that anneals to a specific place in the DNA template, each cycle will result in the synthesis of a single-stranded copy of one strand of the DNA template, resulting in linear accumulation of copies of one of the two DNA strands.

This reaction becomes significantly different in the presence of two different oligonucleotide primers that anneal to opposing strands of the DNA template and direct DNA synthesis towards each other (Figure 10.2). With two opposing primers, the first synthesis cycle results in two progeny DNA molecules with an annealed primer that has been extended by the polymerase past the annealing site for the second primer. Two different types of reaction product occur in all subsequent cycles: primers can anneal to the original two template molecules and to the elongated products synthesized in the preceding synthesis cycle. While copies of the original DNA template accumulate at a linear rate (1, 2, 3, 4, ...), copies of the DNA sequences located between the annealing sites for the two primers accumulate at an exponential rate (1, 2, 4, 8, ...). The region between the primers becomes amplified as a DNA fragment with ends defined by the annealing sites for the primers. When the PCR reaction is allowed to proceed for 20-30 cycles, sufficient yield is obtained to allow the use of gel electrophoresis for observation of the resultant DNA fragment.

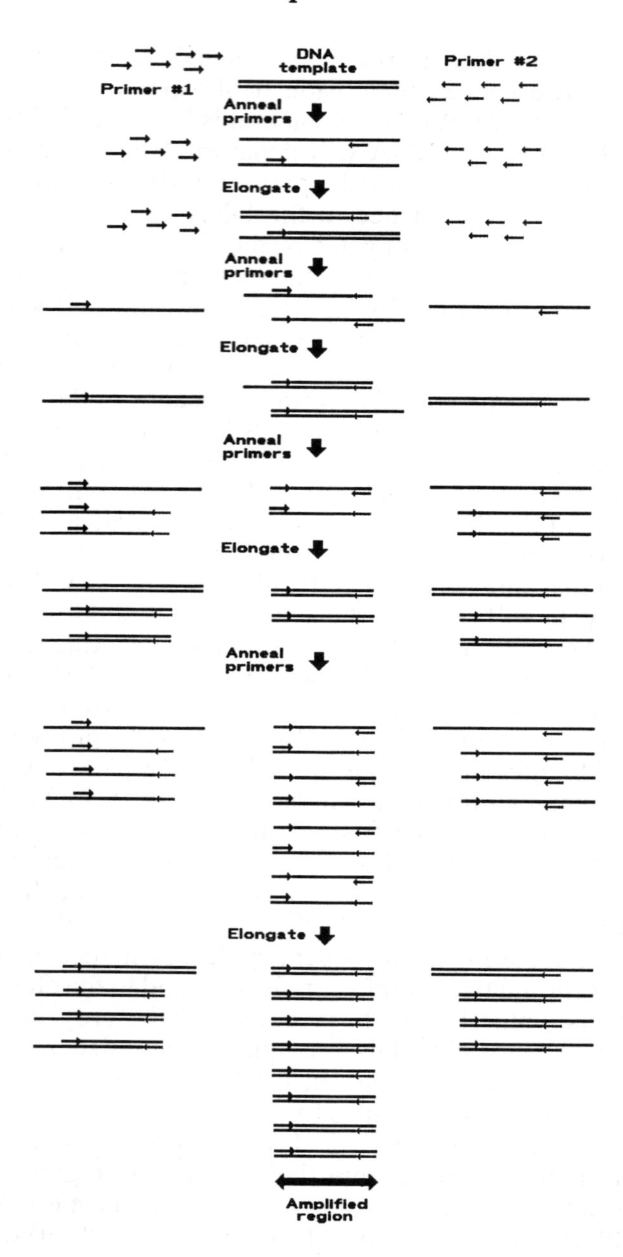
DNA template
Primer #1
Primer #2
Anneal primers
Elongate
Anneal primers
Elongate
Anneal primers
Elongate
Anneal primers
Elongate
Amplified region

The utility of this DNA amplification process is striking. Once a gene fragment has been isolated and sufficient nucleic acid sequence determined to allow the design and chemical synthesis of the oligonucleotide primers, portions of the gene can be rapidly amplified to such high levels that they can be easily examined by gel electrophoresis. Sufficient yields are obtained to circumvent the need for cloning and vector-mediated amplification of DNA fragments in a biological host. Since each amplification cycle can often be completed in a few minutes, the total time required to complete a DNA fragment analysis can be as brief as a few hours, in contrast to the several days required by conventional recombinant DNA methods.

PCR Amplification of RNA (RT/PCR)

The heat-stable DNA polymerases used in the PCR amplification generally work efficiently only with a DNA template and cannot be used to directly amplify RNA molecules. However, the enzyme reverse transcriptase (RT), which can elongate a DNA primer annealed to an RNA template to make a complementary cDNA copy of the RNA molecule, has been combined with PCR to allow amplification of RNA.

The RT/PCR process (Figure 10.3) first anneals an oligonucleotide primer to an RNA sample, then uses reverse transcriptase to elogate the primer and make a cDNA copy of the RNA molecules. The primer may be general, such as oligo dT annealed to the poly-A tail of a eukaryotic mRNA

Figure 10.2. The polymerase chain reaction. The PCR amplification uses two different primers that anneal to opposing sites on the template DNA. The first three cycles of the reaction cause synthesis of a double-stranded DNA fragment that covers the region between the two primer annealing sites. With subsequent cycles, this short fragment accumulates exponentially while all other products accumulate in a linear fashion. The short fragment is amplified and its yield rapidly exceeds that of all other reaction products.

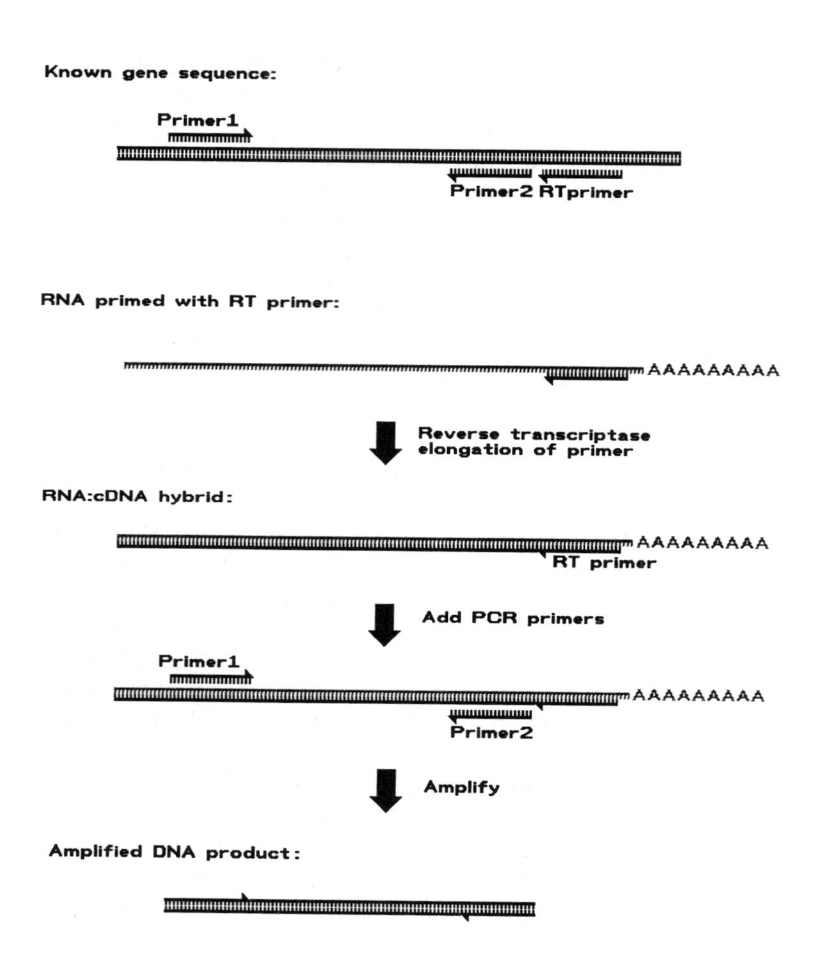

Figure 10.3. RT/PCR amplication of RNA. RT/PCR allows the amplification of an RNA molecule by first annealing an oligonucleotide primer to an RNA sample. Reverse transcriptase is then used to elongate this primer to generate a cDNA population containing the cDNA corresponding to the RNA molecule of interest. In a second reaction step, two specific oligonucleotide primers are added and PCR amplification used to synthesize large amounts of the specific DNA fragment corresponding to the desired RNA molecule.

population, or a specific primer that will anneal preferentially to a particular RNA, such as the rRNA that forms the structural backbone of the ribosome. Following annealing of the primer, reverse transcriptase and deoxynucleoside triphosphates are added to extend the primer and make a single-stranded cDNA copy of the desired RNA molecule(s).

The RT reaction is followed by standard PCR amplification. Two opposing primers specific for the desired gene, a heat-stable DNA polymerase, and deoxynucleoside triphosphates are added to the cDNA and the reaction tube is subjected to at least twenty rounds of enzymatic elongation of annealed primers, heat denaturation of the double-stranded DNA product, and re-annealing of primers to single-stranded DNA templates. As is the case with the standard PCR amplification process, the product of the RT/PCR process is a double-stranded DNA molecule of defined size that can be directly examined by gel electrophoresis.

Strengths and weaknesses of PCR

The PCR process has become an extremely powerful tool in the isolation and restructuring of DNA molecules and, for many applications, has supplanted more conventional recombinant DNA methods. This reaction can often be performed on extremely small quantities of template nucleic acid - the material crudely isolated from a few cells can be sufficient template for PCR amplification of many genes. PCR is much more rapid than conventional gene isolation techniques and, because reactions are typically only 10 to 100 microliters in volume, allows more convenient processing of increased numbers of samples and decreased reagent costs per sample. For many applications, the yields of DNA obtained through PCR amplification are sufficient to circumvent the necessity of cloning in a vector and biological amplification of the desired DNA fragment.

PCR reactions, however, can also be subject to a variety of types of artefact involving the specificity and annealing efficiency of the primers. Manipulation of the sequence of the primers, the ions in the reaction (particularly Mg^{++}), the annealing temperature, and the rate at which reaction conditions change (the ramp times) during the anneal-elongate-denature cycles have all been found to influence efficiency and specificity of the PCR amplification process. While these artefacts can be at times problematic in certain applications, reaction conditions can usually be optimized in a manner that minimizes the production of undesireable amplification products.

As the following examples will illustrate, virtually any nucleic acid manipulation that can be accomplished with conventional recombinant methods can be accomplished more rapidly using PCR technology. Given this observation, one might speculate that PCR technology will eventually entirely supplant recombinant methods. It is important to remember that PCR requires a minimum of two oligonucleotide primers that will anneal to the desired DNA template and allow the amplification process to proceed. Thus, the initial isolation and characterization of a DNA fragment may be dependent on conventional recombinant DNA methods, such as screening of cDNA libraries by nucleic acid hybridization or antibody-based approaches, that do not require nucleotide sequence information.

PCR and RT/PCR are also strongly dependent on the ability of the enzymes to elongate primers both efficiently and with high fidelity through the desired template. While many different RNA and DNA templates have proven ameanable to PCR amplification, researchers occasionally find templates that do not amplify efficiently, presumably due to the inability of the enzymes to extend through particular regions of the nucleotide sequence of the template. The introduction of nucleotide sequence changes during the PCR amplification process has the potential to significantly altering the interpretation of nuleotide sequence information. The introduction of a single base change during the amplification of a cDNA molecule, for example, can cause changes in the amino acid sequence predicted to be encoded by that cDNA, thereby altering predictions about the properties of the protein product. It is for these reasons often important to have as a reference the nucleotide sequence obtained from a DNA fragment that was obtained by conventional recombinant methods without the use of PCR amplification. Although PCR technology is an important development that has greatly reduced the application of conventional recombinant DNA methodology, it appears unlikely that it will completely supplant the older methods.

Detection of alterations of gene structure

One of the most exciting areas in molecular biology involves the attempt to understand how changes in gene structure and function lead to disease, including cancer and various types of genetic disorder. Examination of normal and diseased cells often leads to the hypothesis that change in function of a particular gene is crucial to the disease. Once biochemical studies have implicated a particular gene, one approach to testing the involvement of the gene would be to isolate and compare the genes from normal and diseased cells. A consistent change in gene structure in the disease cells would support the hypothesis that the gene is involved in the pathogenic process.

While the isolation and nucleotide sequence analysis of the gene present in many normal and diseased individuals would be laborious, expensive, and time-consuming by conventional methods, a PCR strategy can be used to simplify the experimental approach (Figure 10.4). The gene or the cDNA corresponding to the RNA product of the normal gene would first be isolated and the nucleotide sequence determined. This established sequence can then be used to design and chemically synthesize PCR primers capable of amplifying the desired gene. Small amounts of nucleic acid would then be isolated from many normal and many diseased individuals and subjected to PCR amplification using the gene-specific oligonucleotide primers. Following nucleotide sequence analysis of each of the amplified DNA fragments, comparison of the sequences would reveal sequence changes that are always associated with the disease.

Many lines of evidence support the notion that many changes in gene function are associated with the occurrence of insertions or deletions of DNA. Since these alterations

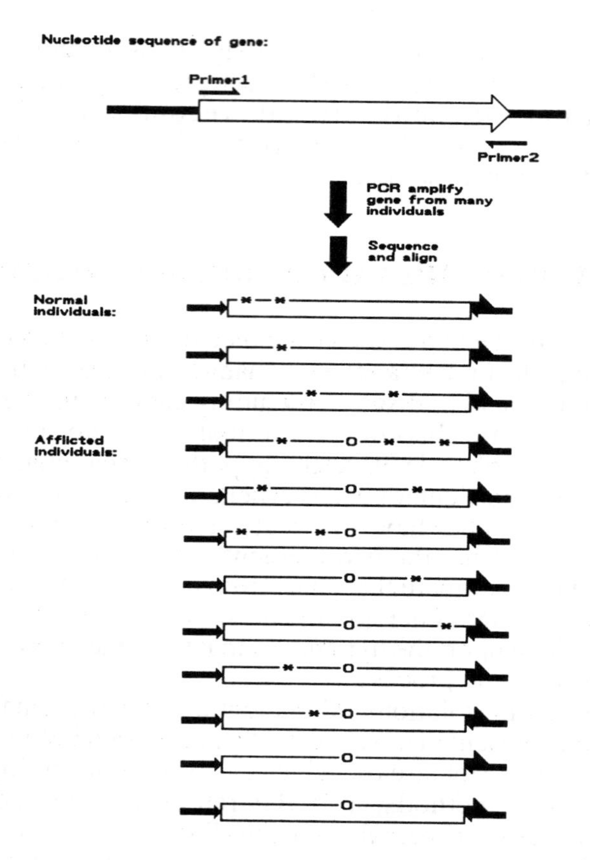

Figure 10.4. PCR strategy to obtain genes for sequence analysis. Once the gene of interest has been isolated and the nucleotide sequence determined, PCR primers can be designed that will amplify all or a specific portion of the gene. These primers can then be used to amplify and determine the nucleotide sequence of this gene in a great number of individuals. Note that because the primers form the ends of the amplified fragments, the sequence of the ends of all of the amplified fragments will be identical to the sequences of the two primers. Individual sequence information is therefore lost at the binding site for each primer, indicated by solid arrows in the illustration. Comparison of the internal sequences obtained from normal individuals and persons afflicted by a genetic disorder may reveal both normal sequence variation caused by the presence of several different alleles of the normal gene (indicated by *) and sequence changes common to those persons who carry the genetic disorder (indicated by o). Those sequence variations common to the afflicted persons are candidates for the cause of the disorder.

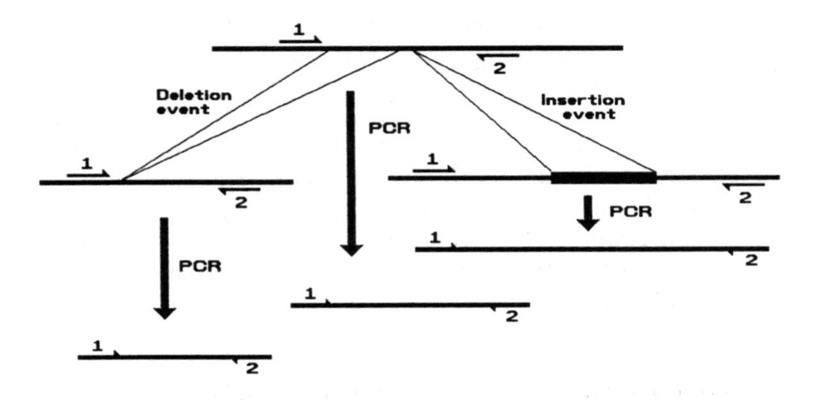

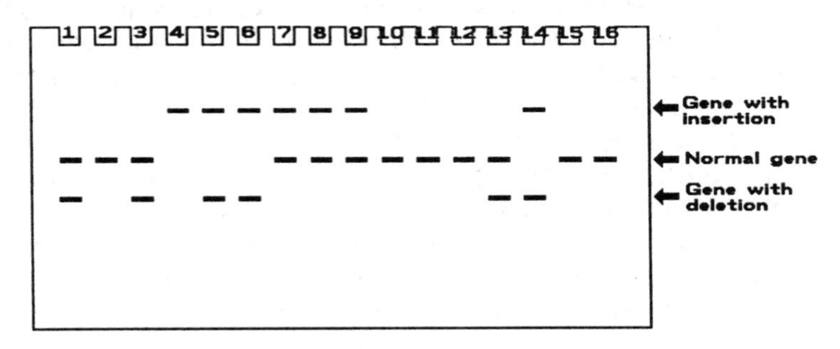

Figure 10.5. PCR used to examine many nucleic acid samples for the presence of deletions or insertions in a gene of interest. Once the gene of interest has been isolated and the nucleotide sequence determined, PCR primers can be designed that will selectively amplify a region of specific interest. The presence of a deletion will decrease the nucleotide distance between the primers, decreasing the fragment size observed following amplification of a gene containing a deletion. The presence of an insertion will increase the nucleotide distance between the primers, increasing the fragment size observed following amplification of a gene containing an insertion. PCR amplification of nucleic acid samples from many individuals followed by gel electrophoresis can reveal the presence of gene copies containing deletions or insertions. PCR can reveal whether an individual is homozygous, containing a single gene type, or is heterozygous, containing two different alleles of the gene. Individuals 2, 10, 11, 12, 15, and 16 are homozygous for the normal gene, and individual 4 is homozygous for the gene with the insertion. All other individuals illustrated were heterozygous and contained two different alleles of the gene.

cause changes in the distance between the annealing sites for a pair of gene-specific PCR primers (Figure 10.5), a PCR-based strategy might use simple gel analysis to examine the size, rather than the nucleotide sequence, of an amplified DNA fragment obtained from normal and diseased individuals. Although the same question can be addressed using conventional blot analysis, the PCR approach requires less material for analysis and allows direct, rather than hybridization-based, visualization of the DNA fragment.

Manipulation of DNA fragments

Although manipulation of isolated DNA fragments can be accomplished by the DNA ligase-mediated addition of oligonucleotide linkers and by various specific modification strategies (see Chapter IX), manipulation is often accomplished more efficiently by the use of PCR. As long as an oligonucleotide primer is long enough to form a stable double-strand hybrid and has sufficient homology between its 3' terminus and the target DNA (generally a minimum of about 4 bases), the primer need not be completely complementary to the intended annealing site. This allows two basic types of modified primer to be used for the PCR-based manipulation of DNA fragments. A primer can contain either internal base subsitutions, or differences relative to the target DNA, that cause changes in the sequence of the amplified DNA fragment, or can contain additional, non-annealing sequences at the 5' terminus that cause the addition of specific sequences to the end of the amplified DNA fragment (Figure 10.6). These types of PCR primer can be used to generate convenient restriction sites for further manipulation of a DNA fragment, or can add important function sequences, such as transcription or translation initiation signals, to an otherwise non-expressable DNA fragment.

PCR can also be used to introduce internal sequence changes in DNA fragments (Figure 10.7). This type of strategy generally requires four PCR primers, two border primers that flank the entire region to be amplified and two

```
a.                          G  G
              5'-CTATACTGCA CT GCTTACGACATCGA-3'
   3'-NNNNNNNNNGATATGACGTGGCTCGAATGCTGTAGCTNNNNNNNNNNNN-5'
```

```
b.

 5'-G
     C
      C
       T
        G
         C
          A
           G
            C
             T
              G
               CTATACTGCACCTAGCTTACGACATCGA-3'
   3'-NNNNNNNNNGATATGACGTGGCTCGAATGCTGTAGCTNNNNNNNNNNNN-5'
```

Figure 10.6. Types of modified PCR primers that can be used to cause a sequence changes at the termini of a DNA fragment during amplification. a) A primer can have an internal mismatch flanked by regions of identity with the target sequence. b) A primer can be identical with the target sequence but possess a completely unrelated sequence at the 5' terminus of the primer. Each of these primers will cause the addition of adjacent *Pst*I (CTGCAG) and *Pvu*II (CAGCTG) cleavage sites to the amplified DNA.

mutant primers complementary to one another and containing the desired internal mutation relative to the original DNA sequence. The mutation is introduced in a two-step process in which the left and right portions of the template are first amplified in two separate primary reactions, one containing the left border primer and the right internal mutant primer and the other containing the right border primer and the left internal mutant primer. The two primary reactions are then mixed, excess left and right border primers is added, and the PCR is continued in a secondary reaction. The two mutant primers added to the

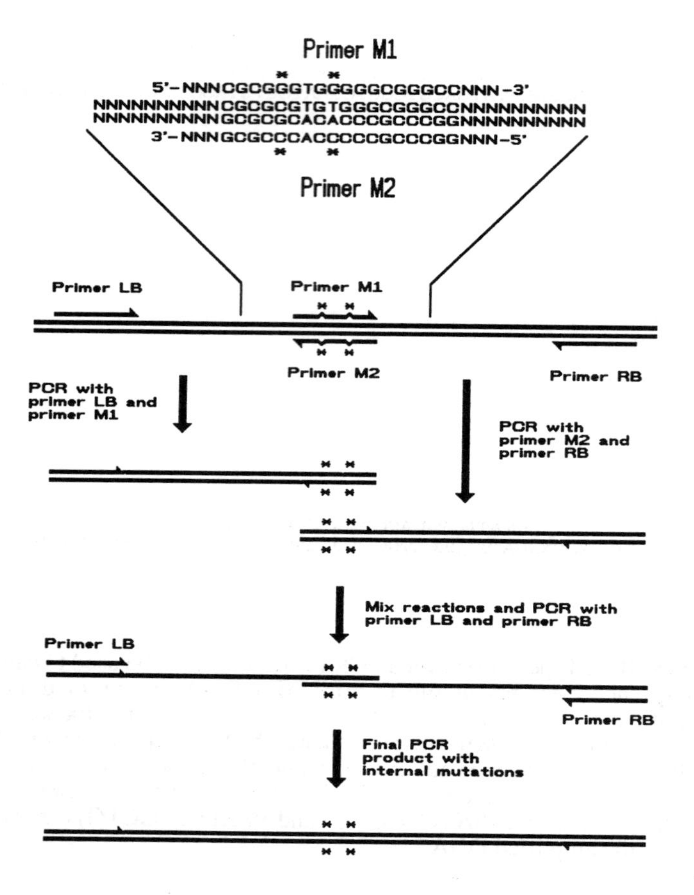

Figure 10.7. PCR used to introduce internal mutations in amplified DNA fragments. The left and right portions of the template are first amplified in two separate reactions. One reaction contains the left border primer (LB) and the right internal mutant primer (M1) and the other contains the right border primer (RB) and the left internal mutant primer (M2). Mutant primers M1 and M2 each contain two base changes relative to the DNA template (*) and are complementary to one another. The resulting left PCR product has a modified right terminus and the right PCR product has a modified left terminus. When these two products are mixed in the presence of excess left and right border primers, the left and right partial amplification products anneal to one another and polymerase extends the 3' termini to generate the full-length DNA fragment containing the desired internal mutation where the mutant primers were incorporated. Subsequent amplification of this full-length, mutated DNA fragment using the border primers causes the accumulation of the desired DNA product.

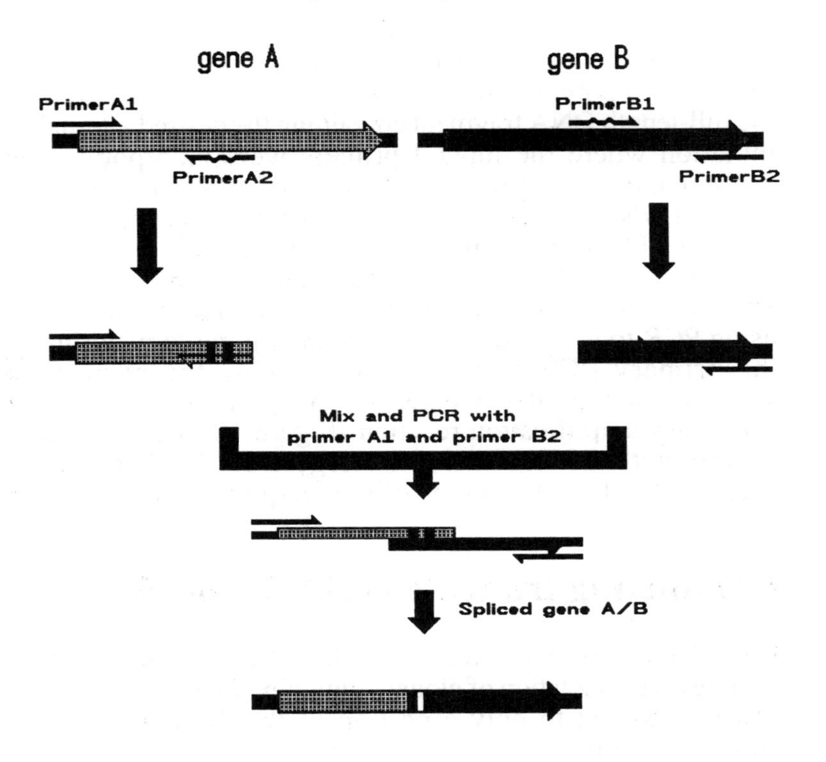

Figure 10.8. PCR used to splice new combinations of gene fragments. Two unrelated gene fragments can be spliced together in a two-step PCR procedure as long as one of the primers used to amplify the desired portion of the first gene has sufficient homology to anneal to one of the primers used to amplify the desired portion of the second gene such that the annealing of the two primers creates the desired gene junction. In the example shown, primer A1 is a perfect match with the sequence of gene A and primer A2 contains several base changes relative to the sequence of gene A. Likewise, primer B2 is a perfect match with the sequence of gene B and primer B1 contains several base changes relative to the sequence of gene B and is complementary to primer A2. In the first step of the PCR splicing process, primers A1 and A2 are used to amplify the left portion of gene A and primers B1 and B2 are used to amplify the right portion of gene B. When the primary PCR products are mixed and PCR amplification continued in the presence of primers A1 and B2, the left portion of gene A can anneal with the right portion of gene B through the homology present at the site of incorporation of primers A2 and B1. Extension by DNA polymerase generates the desired spliced gene A/B product and subsequent amplification by primers A1 and B2 generates usable yields of the product.

left and right partial amplification products anneal to one another and polymerase extends the 3' termini to generate the full-length DNA fragment containing the desired internal mutation where the mutant primers were incorporated. Subsequent cycles of amplification of this full-length, mutated DNA fragment using the border primers result in usable yields of the desired DNA product.

This strategy can be modified to allow the *in vitro* splicing of otherwise unrelated DNA fragments by first using PCR to introduce a short region of homology during the primary PCR reaction, then mixing the primary amplification products together and joining them in a secondary amplification reaction (Figure 10.8). Although feasible with conventional technology, PCR facilitates the speed and utility of this type of *in vitro* gene splicing.

Completing truncated cDNA clones

One of the common artefacts in the screening of cDNA libraries is the isolation of clones containing DNA fragments that correspond to only a portion of the desired cDNA coding region. General cDNA libraries screened by hybridization strategies often yield clones that are missing the 5' terminus, while expression libraries screened with antibodies may yield clones that contain only an internal fragment of the desired gene. Once the nucleotide sequence of the partial clone has been determined, PCR can be easily used to obtain the missing regions of the gene.

The procedure, often termed Rapid Amplification of cDNA Ends (RACE) (Figure 10.9), makes use of the known sequence of the gene fragment to design and synthesize two oligonucleotide primers. In the 5' RACE, one oligonucleotide, the RT primer, is designed to anneal to the desired RNA molecule and be extended by reverse transcriptase to make a single-stranded cDNA copy of the missing portion of the RNA molecule. Following the reverse transcriptase extension reaction, a short stretch of a defined nucleotide sequence is added to the 3' terminus of the newly synthesized cDNA by either ligation of an oligonucleotide or terminal transferase-mediated addition of a polynucleotide tail, such as poly-dA.

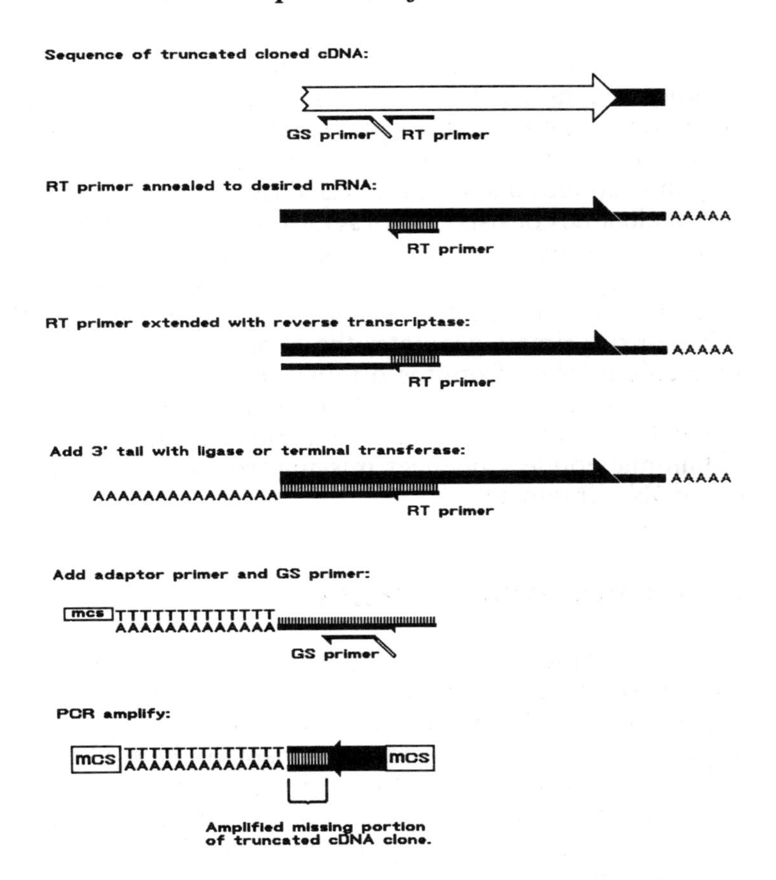

Figure 10.9. RACE to obtain missing ends of cDNA fragments. The nucleotide sequence of the cloned, truncated cDNA is used to design two oligonucleotide primers, the RT primer and the gene-specific GS primer, that will anneal to the desired RNA molecule and direct synthesis into the missing region of sequence. The RT primer is then annealed to an RNA sample and reverse transcriptase used to synthesize a cDNA copy of the desired region of the RNA molecule. A defined oligonucleotide sequence, such as oligo-dA, is added to the 3' terminus of the newly synthesized cDNA to provide a binding site for a PCR primer. The tailed cDNA is then used as template in a PCR reaction containing a tail-specific adaptor primer and the GS primer, which will both amplify and add convenient restriction endonuclease cleavage sites (mcs) to the PCR product. The amplified PCR product can be directly sequenced or digested with endonucleases that cleave within the sites present in the mcs portion of the primers and cloned in a standard vector for further analysis.

Two new oligonucleotides are then added to the newly synthesized, tailed cDNA. One is a primer consisting of a 5' region containing one or more restriction cleavage sites and a 3' region capable of annealing to the synthetic oligonucleotide added to the 3' cDNA terminus. The second olignucleotide consists of a 5' region containing one or more restriction cleavage sites and a 3' region capable of annealing specifically to the newly synthesized cDNA adjacent to the annealing site for the RT primer. PCR amplification with this pair of primers generates the 5' portion of the cDNA molecule missing from the original, partial-length cDNA clone, allowing completion of the desired, full-length cDNA nucleotide sequence. Similar strategies can be applied to obtain nucleotide sequences missing from the 3' terminus of a cDNA fragment.

Obtaining related genes

The requirement for knowledge of some nucleotide sequence of the desired gene to allow design of PCR primers can at times even be circumvented. The isolation, nucleotide sequence determination, and comparison of the equivalent gene from several different organisms has for numerous genes revealed that genes that code for similar protein activities generally have very similar nucleotide sequences. This observation may simply illustrate that, given the limited number of amino acids that are used to assemble proteins, two proteins that have similar functions must also have similar amino acid sequences. Since the amino acid sequence of a protein is encoded by the sequence of nucleotides in the gene, similar proteins will likely be encoded by similar nucleotide sequences. This principle has been illustrated in Figure 10.10, where the hypothetical coding regions for a gene isolated from rat, mouse, hamster, goat, and catfish have been aligned and areas where the sequences are identical to the rat sequence presented as solid lines.

Once the nucleotide sequences of two or more closely related genes have been determined, alignment can reveal regions with a high degree of nucleotide sequence identity.

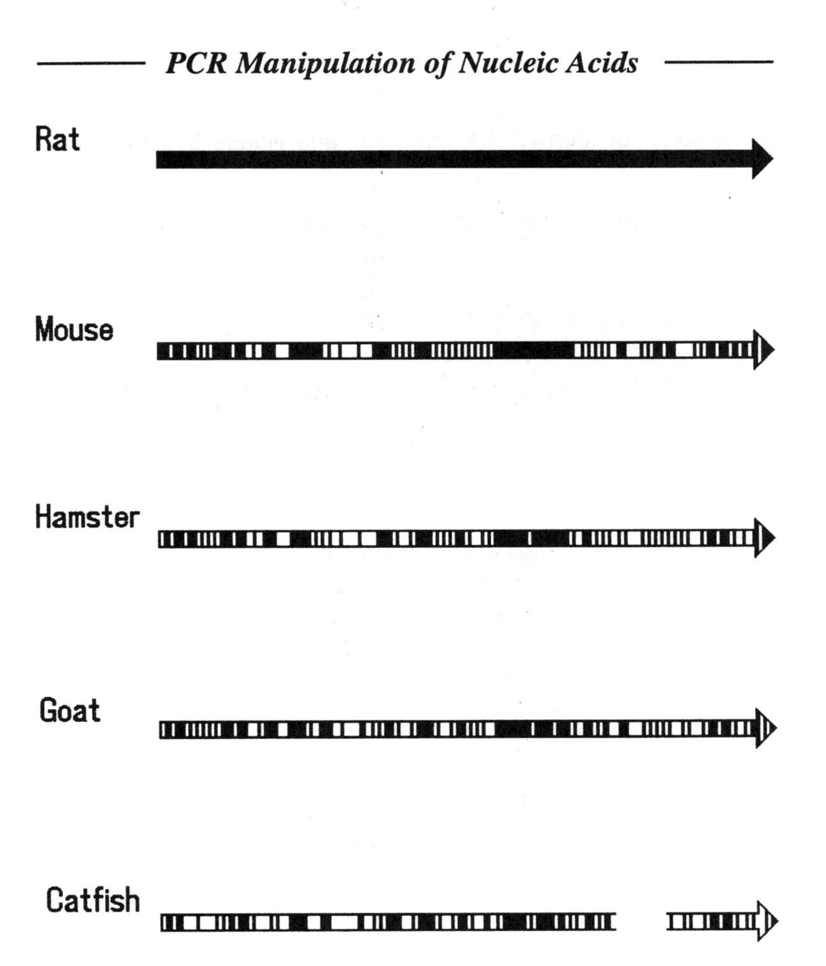

Figure 10.10. Nucleotide sequence homology among related genes. Examination of the nucleotide sequences of genes that encode similar proteins in different organisms has revealed that the genes often have very similar nucleotide sequences. This principle is illustrated by comparing the nucleotide sequences of the equivalent gene isolated from mouse, hamster, goat, and catfish to the gene isolated from rat. A solid bar indicates a region where nucleotide sequence is identical to that observed in the rat gene, while an open bar indicates a region where the nucleotide sequence is different than that observed in the rat gene. A region of sequence that appears to be absent from the catfish gene is indicated by a break in the aligned catfish sequence. It is a common observation that organisms that appear similar to one another (such as rat and mouse) have a higher level of sequence identity than do organisms that appear quite different from one another (such as rat and catfish).

Regions of identity between rat and mouse sequences:

Potential human gene primer pairs:

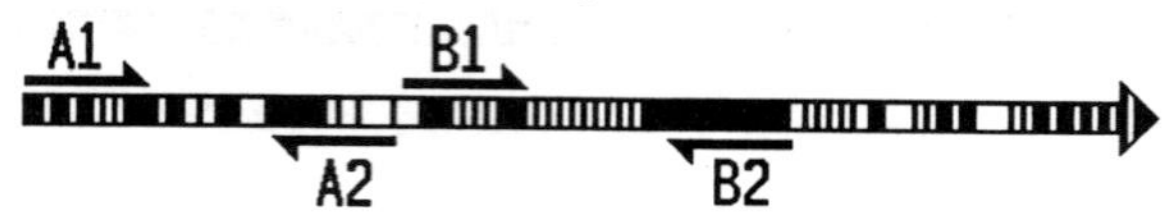

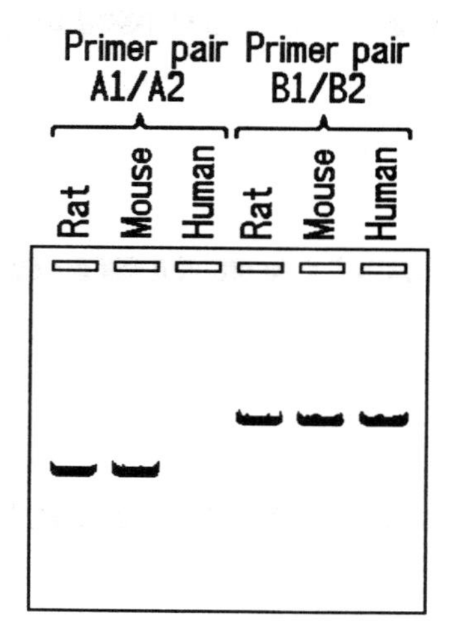

Figure 10.11. Amplifying genes using PCR primers designed based on nucleotide sequences obtained from different organisms. The nucleotide sequences of a gene obtained from both rat and mouse were aligned with regions of identical nucleotide sequence indicated by solid bars and nonidentical sequence indicated by thin lines. Two sets of PCR primers, A1/A2 and B1/B2, were chosen that occur in regions of complete identity between the rat and mouse sequences in the hope that these primers would allow amplification of the human copy of the gene. When these primers were used to amplify rat, mouse, and human DNA and the PCR products examined by gel electrophoresis, the results revealed the anticipated DNA fragment sizes for both the rat and mouse DNA samples. The human DNA sample, however, yielded a PCR product of the anticipated size only for primer pair B1/B2, suggesting that these primers could be used to amplify and isolate a portion of the human gene for further sequence analysis.

These regions can be used to design potential PCR primers to amplify the equivalent gene from an organism from which no information is yet known regarding the nucleotide sequence of the desired gene (Figure 10.11). These primers are merely guesses at what the nucleotide sequence in the desired gene might be based on the appearance of highly conserved sequences in closely related genes. While not all pairs of PCR primers designed using this strategy will successfully amplify the desired genes from unrelated species, primer combinations can be found that will successfully amplify the same gene from many different species or will amplify many closely related, yet different genes from within a single organism.

Summary

1. **The Polymerase Chain Reaction (PCR) is a very powerful DNA method that uses a heat-stable DNA polymerase to amplify DNA or RNA from very small amounts of starting material.**

2. **PCR requires some information regarding the nucleotide sequence of the target DNA to allow the design and synthesis of oligonucleotide primers that direct the amplification of the desired target gene.**

3. **The speed, convenience, and decreased costs associated with PCR have helped to supplant some of the more conventional methods previously used to isolate, characterize, and manipulate nucleic acids.**

11

The Application of Molecular Biology

"It is not necessary to understand things in order to argue about them." — Pierre De Beaumarchais

The methods called recombinant DNA developed out of the scientific desire to address the fundamental hypothesis that the physical traits of a cell are encoded by the nucleic acids present in the chromosomes of the cell and changes in the structure and function of nucleic acid result in changes in cellular characteristics.

How has this new methodology succeeded in application to testing of fundamental scientific theories? Application of these methods to the study of developmental biology has provided significant support for the **cell theory**, which suggests that all organisms are composed of many individual cells, by allowing the investigation of the changes in gene expression that accompany the development of a mature organism. The ability to isolate, modify, and re-introduce specific genes has been instrumental in providing overwhelming evidence in support of the **chromosomal theory of heredity**, which suggests that the chromosomes within the individual cells control the physical traits of the cells. The isolation and sequence analysis of related genes from several different species and the study of the occurrence of alleles in wild populations have provided information that supports elements of the **theory of evolution by natural selection**, which suggests that complex organisms are derived from more primitive organisms by a process of accumulation of changes in physical traits of cells.

Development and continued refinement of the methods of molecular biology have been critical in the explosion of information regarding the role of nucleic acids in the control of the physical traits of cells. The increased understanding of the structure and function of the genome has also resulted in the appearance of molecular biology in everyday life.

It is important to understand that scientific research can be divided into two general categories: basic and applied research. Basic research might be described as the search for a basic understanding of a problem area, while applied research can be thought of as applying the results of basic research to generate a useful product.

The scientific methods and the research topics that make up the field of molecular biology are obviously of interest to the scientific community involved in basic research, but why should these topics generate widespread interest and controversy among people who are not members of the scientific community? The increasing use of molecular biology methodology in applied research has been accompanied by the appearance of biomedical and production innovations that are of increasing public notice.

Understanding of scientific theory is rarely necessary for everyday application of a scientific principle. After all, a person need not understand either electricity or the internal combustion engine to be fairly confident that an automobile with a full tank of gas and a good battery will start when the key is turned in the ignition. It is not really necessary to understand the **mechanics** of technology to be able to use the **products** of technology. Consider that the vast majority of people in the United States understand how to use fire to cook and heat, yet very few of these people can start a fire in the absence of technological devices (matches or lighters, for example).

Recombinant DNA technology developed out of the desire to obtain large amounts of a purified gene to address various important hypotheses regarding gene structure and function. The concepts that allow recombinant DNA technology are fairly simple. Restriction endonucleases are proteins that cleave DNA in a site-specific manner to generate specific, reproducible fragments of the DNA present in a chromosome. DNA fragments can be separated according to size by gel electrophoresis and stained with dyes that allow the detection of as little as 5×10^{-9} gram of DNA. Digestion of a DNA fragment with several different enzymes can be used to establish a physical map of the positions of cleavage sites relative to one another - a restriction map of the DNA fragment. Restriction maps can provide direct correlation between the physical structure of a chromosome and the order of the genes present on the chromosomal DNA molecule.

DNA fragments can be inserted into cloning vectors designed to propagate in a suitable host. The vector serves two functions: a fragment inserted into the vector has been molecularly cloned away from other DNA fragments and once cloned, can be amplified by purification of the recombinant molecule. Large amounts of a very pure DNA fragment can be obtained by cloning the fragment in a vector. Molecular cloning of DNA fragments overcomes many of the problems associated with the isolation of rare DNA molecules and the preparation of sufficient amounts of DNA for many experiments.

As discussed in the previous chapter, these methods are of tremendous value in the analysis of the biological mechanisms that are involved in regulating gene expression and have contributed to a virtual revolution in molecular biology. The new information accumulates a rate that causes a continuing re-evaluation of scientific theories regarding gene structure and function. The scientific debate that arises is part of the normal process of scientific evaluation. Experiments are designed to test theories, results are interpreted, and the results are compared with existing theories to determine whether the theories are consistent with the experimental results. Controversy is accepted among scientists as a normal part of the investigation process.

Recombinant DNA methods and their application to molecular biology problems have generated not only scientific debate, controversy, and re-evaluation of theories regarding gene structure and function, but have also generated significant public debate regarding safety, ethics, and economic potential of this type of research. These methods have been perceived as having tremendously powerful positive and negative social impacts, with both benefactors and detractors making misleading statements. Understanding how these methods can be applied to various specific problems can help illustrate why recombinant DNA methodology is no longer merely an investigative tool for molecular biologists, but is experiencing increasing application to common problems.

Contributions to genetic screening and analysis

Molecular biology has provided several new tools, including restriction fragment length polymorphism (RFLP) analysis, the polymerase chain reaction (PCR), and DNA fingerprinting, that are becoming increasingly common in the determination of genetic traits of individuals. These methods of analysis are all based on the principles that since the DNA of an organism contains the genes that code for all of the physical characteristics of a specific individual and since each individual (with the exception of identical twins) has a unique combination of genes, a DNA sample isolated from an individual can be used to predict physical traits or to actually identify the individual out of a group.

Restriction fragment length polymorphism and RFLP analysis

RFLP analysis is based on the observation that there are slight differences in the nucleotide sequences of the same gene isolated from two different individuals. These nucleotide sequence differences can be used to distinguish the gene isolated from one individual from the same gene isolated from a different individual. Isolation and determination of the nucleotide sequences of the genes from both individuals would be a slow and expensive method of identification. Fortunately, the variation in nucleotide sequence will occassionally either inactivate or create the cleavage site for a restriction enzyme (Figure 11.1). When the DNA from two individuals is digested with the same restriction enzyme, the same gene can be present on DNA fragments of different sizes. The lengths of the restriction fragments carrying a specific gene are polymorphic, or of variable sizes, when an entire population of indiviuals is examined.

Restriction fragment length variability can be examined by Southern blot analysis using a labeled probe for a specific gene to identify different fragments that carry slightly different copies of the same gene. Eukaryotes are generally diploid, or contain pairs of chromosomes, and a given gene will be located at the same position of a specific chromosome. Each eukaryotic cell will therefore contain two alleles or copies of the same gene (one on each of two copies of each chromosome). Since the sperm and egg are haploid and contain one rather than two copies of each chromosome, one chromosome in the fertilized diploid zygote comes from each of the two parents. One copy of each gene is maternal, or derived from the mother, and the other copy is paternal, or derived from the father. If the maternal and paternal genes were slightly different, an individual will carry two slightly different copies of the same gene. When these different alleles are associated with different restriction fragment sizes, the restriction fragments can be used as markers to track the alleles as they are passed from parents to progeny (Figure 11.2).

Medical research is identifying an increasing number of polymorphisms that are associated with specific genetic disorders. These polymorphisms may affect the gene involved in the disorder or may be simply associated with DNA that is very close to the defective gene. When association of a specific restriction fragment size with a particular genetic disorder is very high, that fragment can serve as a genetic marker for the disorder and can be used to screen for individuals who are carriers or victims of the disorder (Figure 11.3). With a recessive trait, heterozygous individuals with one copy of the defective gene will be carriers with a 50% probability of transmitting the defective

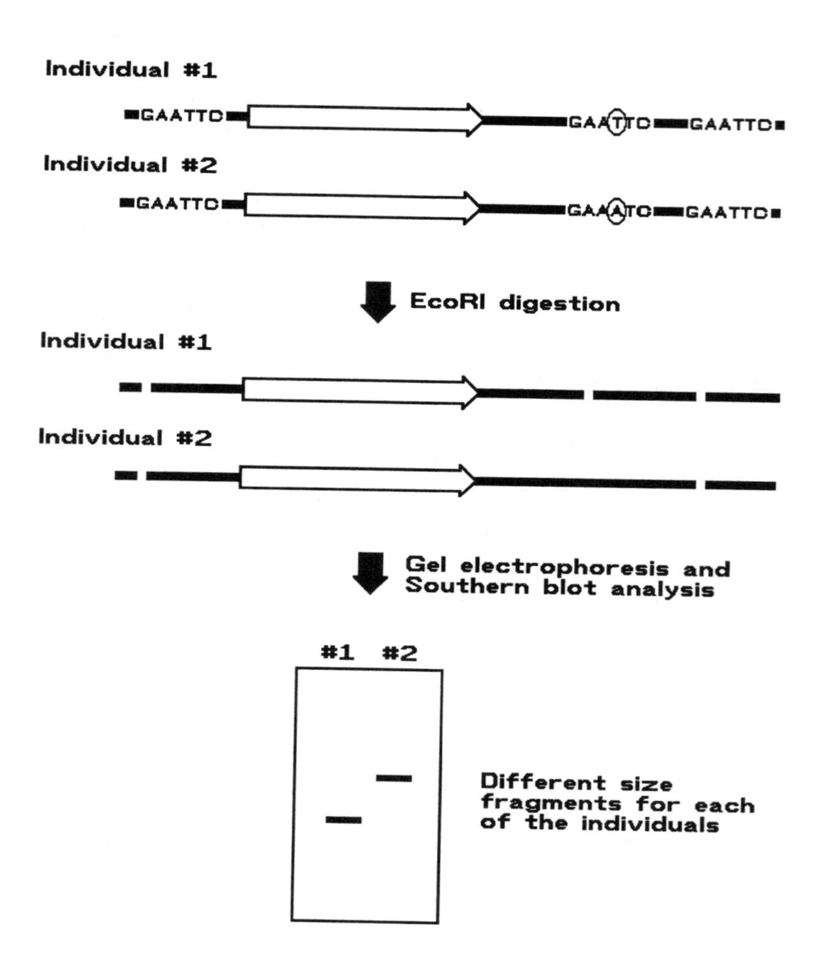

Figure 11.1. Minor variations in nucleotide sequence can lead to differences in the presence of restriction enzyme cleavage sites in different chromosomes. The same gene may be present on different size DNA fragments in different individuals. In the example illustrated, the nucleotide sequences of two DNA fragments containing the same gene differ at only a single nucleotide residue, where a T in individual #1 is replaced with an A in individual #2 (residues are circled). This single base change inactivates an *Eco*RI cleavage site in individual #2 (GAATTC changed to GAAATC). The same gene will be present on a different size *Eco*RI fragment in these two individuals.

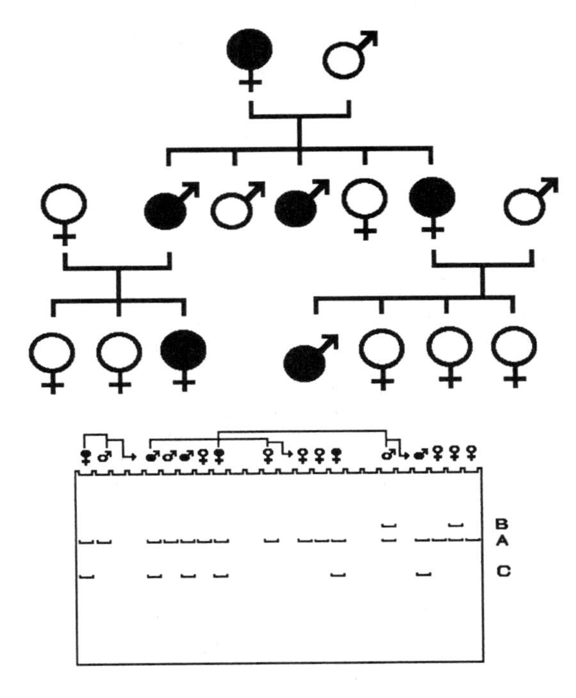

Figure 11.2. A family tree that traces the transmission of a genetic disorder from parents to children can sometimes be matched with a restriction fragment length polymorphism. In the example tree shown, normal individuals are indicated by open symbols and affected individuals by filled symbols. This pedigree traces transmission of a genetic disorder from an affected mother and unaffected father through their five children and seven grandchildren. When DNA from each individual is isolated and subjected to Southern blot analysis with a DNA probe for the affected gene, all samples should contain two copies of the gene examined (one on each of the pair of chromosomes that carry the gene). All samples examined contained a common band (A), which evidently corresponds to a normal copy of the gene. The unaffected male spouse of one child also contained a different band (B), which was transmitted to an unaffected grandchild. Band B corresponds to a normal copy of the gene with a polymorphism that changes the size of the DNA fragment without affecting gene function. In addition to the presence of band A, indicating a normal copy of the gene, all of the affected individuals contained a smaller band (C). Band C serves as a physical marker that indicates the presence of the defective gene. Since the affected individuals contained a normal copy of the gene in addition to the defective copy of the gene, the defective gene must be dominant, controlling phenotype in the presence of the normal gene.

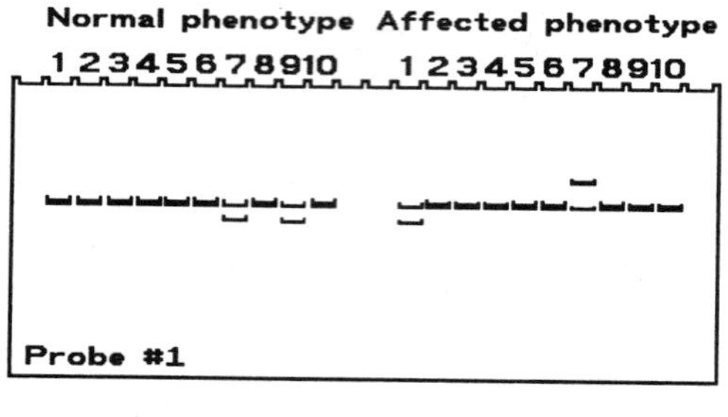

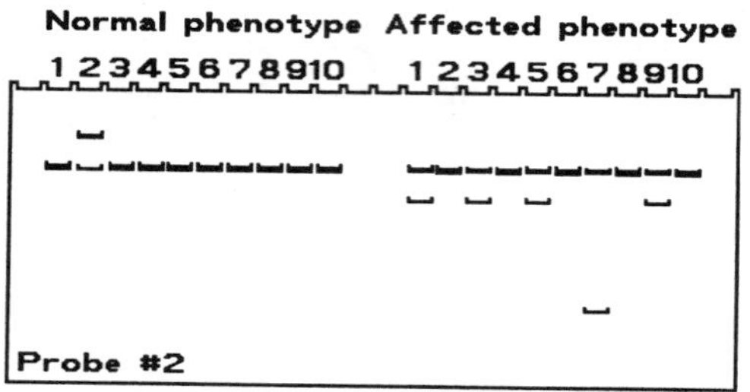

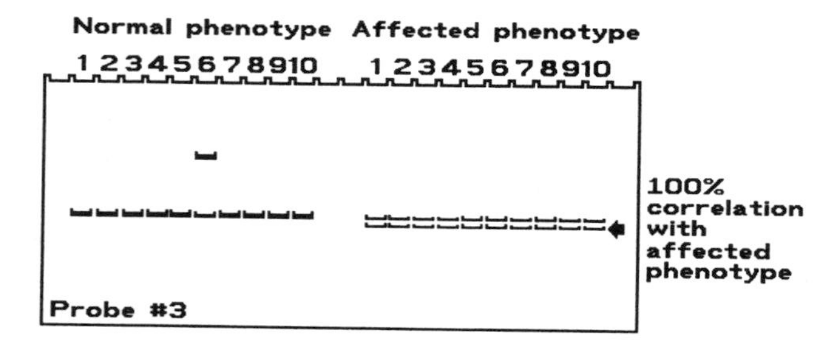

Figure 11.3. Choosing a restriction fragment length polymorphism as a marker for a genetic disorder involves finding a DNA probe that exhibits 100% correlation with the affected phenotype. In the illustration, the DNA of ten normal and ten affected individuals has been cleaved with a restriction enzyme and examined by Southern blot analysis with three different DNA probes. The fragments revealed with probe #1 show only normal polymorphism - two normal and two affected individuals have different fragments. Results obtained with probe #2 reveal that a fragment present in four affected individuals does not occur in any normal individuals, but six of the affected individuals examined do not contain this fragment. While probe #2 may be against a DNA fragment located close to the affected gene, it does not include the gene itself. Probe #3 reveals a DNA fragment that has 100% correlation with the disease phenotype - all of the affected individuals and none of the unaffected individuals contain this fragment. This probe must be located very close to and may include the affected gene.

gene to progeny and homozygous individuals with two defective alleles will have a 100% probability of transmitting the defective gene. RFLP analysis allows genetic counselors to examine the DNA of individuals who have a family history of a genetic disorder and determine the probability of transmitting a genetic disease to children, helping people make informed decisions about the risk of having children with genetically transmitted disorders.

PCR and RFLP analysis

RFLP analysis generally requires several micrograms of DNA for analysis, an amount easily obtained from a small tissue or blood sample. The technique of polymerase chain reaction (PCR) amplification of DNA is a development that has greatly reduced the amount of DNA necessary to perform genetic screening. Unlike conventional RFLP analysis, PCR analysis can be accomplished with extremely small amounts of DNA. The few cells that remain at the end of a human hair or in a drop of blood, for example, can provide sufficient DNA for the PCR amplification and analysis of specific DNA sequences. The PCR procedure requires more initial information than RFLP analysis, since the nucleotide sequence of the DNA fragment to be amplified must first be determined to allow the design of primers that will flank the region and direct the amplification process. PCR amplification can be combined with Southern blot analysis to allow sensitive examination of specific sequences in extremely small samples of DNA.

DNA fingerprints

The principles of RFLP and PCR analysis are frequently applied in a form of DNA characterization called DNA fingerprint analysis. Although the coding regions of most genes show relatively little sequence variability, some regions of the chromosomes have extreme sequence variation. The human genome contains short nucleotide sequence elements called satellite DNA that are present in thousands of copies per cell. Many adjacent, repeated copies of a short sequence are present at specific locations in mammalian chromosomes, but the precise number of the repeated copies can be quite variable, causing slight variations in the total size of the region containing the repeated element (Figure 11.4). The DNA fragments containing this repetitive DNA can be of slightly different sizes in different individuals. Analysis of these repeated elements with either RFLP or PCR methods can lead to a pattern of DNA bands, often called a DNA fingerprint, that can be used to identify the individual from whom the DNA sample was purified.

Matching a DNA fingerprint with a DNA sample is based on the experimental observation that although a region of DNA that contains a repeated element can have variation in the number of copies of the repeat, some copy numbers occur more frequently than other numbers. When DNA samples isolated from members of the population, a DNA fragment size associated with a particular repeated element might occur with a frequency of 12 times out of 100 individuals (12/100) examined. A different fragment size associated with the same repeated sequence might occur much less frequently and be found in only 1/100 individuals examined.

Since people are diploid and have two copies of each chromosome, the DNA fingerprint pattern of an individual will contain the pattern of bands on the chromosome received from the mother **plus** the pattern of bands on the chromosome received from the father. The overall chances that a DNA sample contains a specific combination of DNA fragments can be expressed by multiplying together the probabilities observed for each of the two chromosomal patterns. The probability that a given individual will have both the band found in 12/100 people and the band found in 1/100 people is 12/100 X 1/100 = 12/10,000 or 0.12%. It would be necessary to examine about 833 different people to find someone who by coincidence alone has the identical pattern of DNA bands. Whether two DNA samples with the same DNA fingerprint are identical can be expressed in terms of the probability of randomly finding two unrelated individuals with the same band patterns.

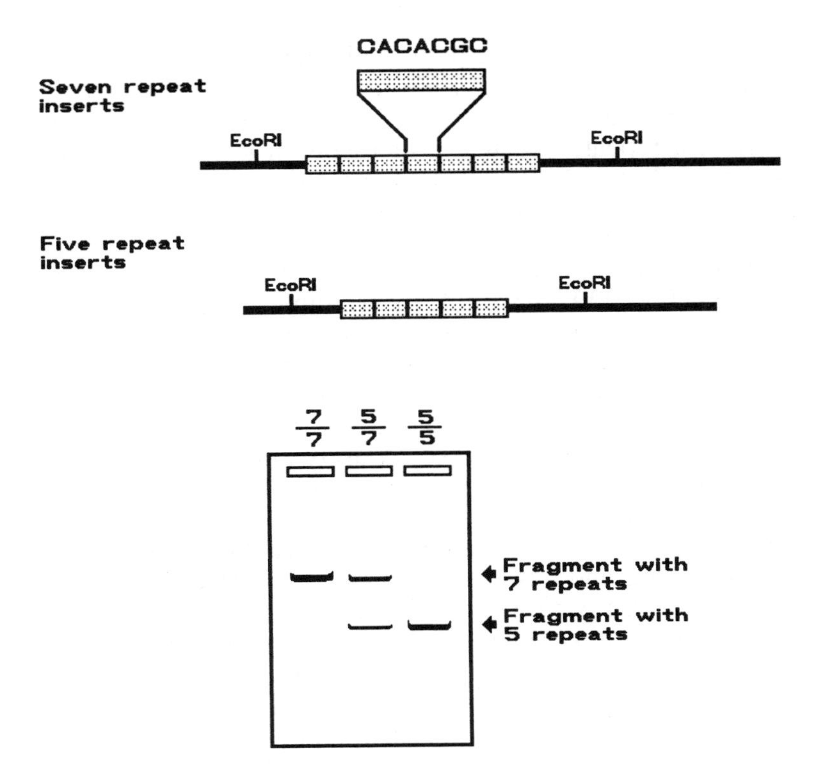

Figure 11.4. Short, linearly repeated sequence elements like the sequence CACACGC can be present at different copy numbers in the same region of different chromosomes. Although each member of a pair of chromosomes contains the same genes, one chromosome may contain seven copies of the repeated sequence at a specific location while the other chromosome of the pair has five copies of the repeated sequence. When the two DNA molecules are digested with a restriction enzyme that does not cut within the inserts, like *Eco*RI, one fragment will be larger than the other. In the example illustrated, one fragment would be two inserts, or 14 bases, larger than the other fragment. Since eukaryotic cells contain pairs of chromosomes, a human cell could be homozygous and have a single DNA band indicating two chromosomes with seven repeat inserts each (7/7), homozygous and have a single DNA band indicating two chromosomes with five repeat inserts each (5/5), or heterozygous and have two different bands indicating one chromosome with five repeat inserts and one with seven repeat inserts (5/7).

By using DNA sequence elements that have been demonstrated to have a wide range of variability and by examining the DNA patterns corresponding to several different repeated sequence elements, it is easy to obtain calculated probabilities that are astronomical. If two DNA samples both contain four different bands that occur at a frequency of 1/200 individuals, the total probability of this fragment combination is 1/200 x 1/200 x 1/200 x 1/200 = 1/1,600,000,000. It would, in principle, be necessary to examine 1,600,000,000 people to find two people that coincidentally have this DNA fingerprint pattern. For sufficiently rare fragment sizes, the calculations can suggest that there are not enough people on the planet to find an identical DNA fingerprint on a purely random basis.

The DNA fingerprint has become a molecular biology tool with increasing legal significance. The DNA fingerprint can be much more conclusive than conventional blood and tissue typing and has been used to decide cases involving alleged paternity of a child. Since a child receives half of its chromosomes from the mother and half from the father, all of the bands in the DNA fingerprint of a baby should match DNA bands present in the DNA fingerprints of either the mother or the father. In principle, bands present in the fingerprint of the baby but absent from the fingerprint of the mother must be present in the fingerprint of the alleged father (Figure 11.5), or the alleged father **cannot** be the biological parent of the child.

DNA fingerprinting is a tool with tremendous potential for forensic investigation. A DNA fingerprint can be performed on the DNA extracted from a small amount of tissue left by the perpetrator at the scene of a crime. Some crimes, such as rape, leave tissue samples that allow very reliable DNA fingerprint analysis. PCR technology is so sensitive that a reliable DNA fingerprint can be obtained from the DNA isolated from a drop of blood dried on a victim's clothing, from the skin cells that lodge under a victim's fingernails when scratching an attacker, or from a hair found at the crime scene. By comparing the DNA fingerprints obtained from these small samples that are often found at the scene of a crime with the DNA fingerprints of principle suspects, it is possible to obtain statistics that either eliminate suspicion of involvement or further implicate a suspect.

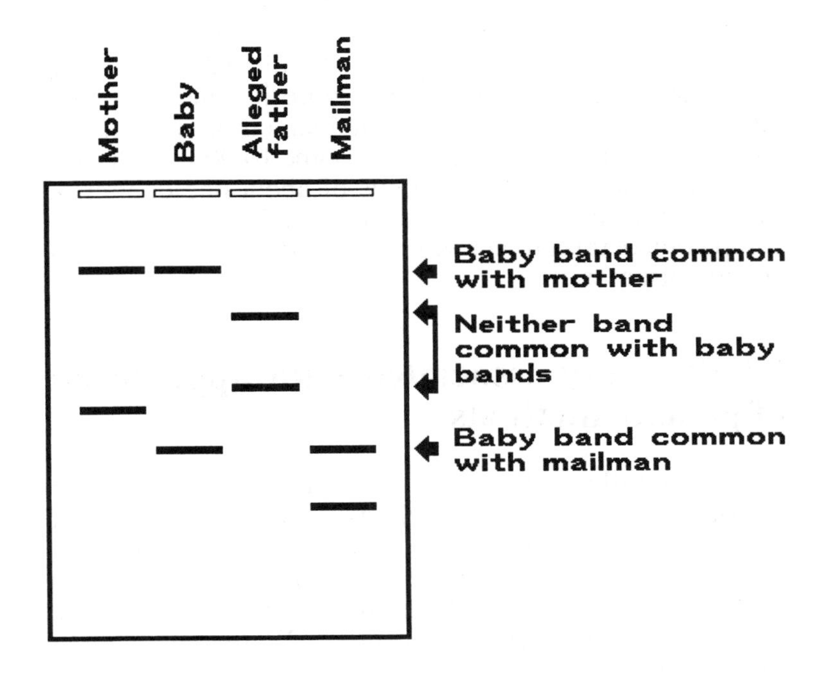

Figure 11.5. Use of DNA fingerprints to resolve a case of disputed paternity. The mother of a child accuses a man of being the father of her baby and demands child support payments. The alleged father denies the child is his and demands a DNA fingerprint to support his claims. A DNA fingerprint test including DNA isolated from the mother, the baby, the alleged father, and the mailman (included at the recommendation of the alleged father) was prepared. One band observed in the DNA fingerprint of the baby was clearly present in the DNA of the mother, as expected, but the other band present in the baby was not present in the alleged father and neither of the bands observed in the alleged father was present in the baby. Since the bands present in the baby must come from either the mother or the father, these results suggest that the alleged father is not the biological father of the baby. Curiously, the non-maternal band present in the fingerprint of the baby was the same size as one of the bands in the DNA fingerprint of the mailman. This finding does not prove that the mailman is the father of the child (the matching band may be extremely common), but is consistent with this suggestion.

The Human Genome Project

The human genome is believed to contain some 80,000 to 100,000 different genes. The human genome project is an attempt to create a chromosomal genetic map of the locations of all of these genes. This project, which involves the efforts of hundreds of scientists all over the world, involves restriction mapping, genetic analysis, and nucleotide sequence analysis of human chromosomes. To date, the chromosomal locations of some 16,000 genes have been identified and the functions of about one quarter of these genes have been determined. It is hoped that completion of this project will facilitate the analysis of many genetic diseases.

Controversy associated with application of genetic methods

Relatively little controversy has been associated with the use of molecular biology methods to help couples at risk for genetic disorders decide whether to have children. However, these methods have also significantly increased the ability of physicians to diagnose a variety of fetal genetic disorders following an amniocentesis. Women who are at risk for transmitting a genetic disorder to their children can have the chromosomes of the unborn fetus examined and learn, with a very high degree of probability, whether the child is destined to be a victim of the genetic disorder. A pregancy might then be terminated to avoid giving birth to a genetically doomed child.

Advocates of this genetic technology might argue that the procedures can prevent the unnecessary suffering of both parent and child by allowing selective abortion based on an informed decision about the presence of a genetic disease. Opponents have pointed out that this application represents the first step towards eugenics - the early termination of pregnancy to avoid unwanted characteristics or to select for desired traits in offspring. Where should lines be drawn? If the method is allowed to provide genetic information as justification for abortion, which types of genetic result justify termination of pregnancy? Is the presence of a genetic disease that will result in severe retardation and death by the age of 10 years of the same consequence as the finding that a fetus is male in a family of ten boys and no girls? This is not really a scientific question, but a dilemma that has arisen because of the rapid development and application of scientific methods.

The application of DNA fingerprinting, which is theoretically sound and scientifically reliable, has also been accompanied by controversy. Scientific innovations naturally move from the realm of research facilities to the domain of companies that provide service or product in exchange for a fee. In the case of DNA fingerprinting for legal identification of individuals, some concern has arisen over both the reliability of the DNA fingerprinting process and the capability of the consumer (the legel system) to evaulate the significance of the results.

In one case where a suspect was convicted of murder, the jury was described to have decided "you can't argue with science" when confronted with the DNA fingerprint evidence. A significant number of molecular biologists who subsequently examined the same DNA fingerprint data came to a conclusion opposite that presented by the company that performed the tests and convinced the jury! Subsequent testing procedures have revealed levels of error that would be deemed unacceptable for scientific investigations. Although the scientific method might be sound, the quality of the test is limited by the reliability of the individual performing the test and interpreting the data.

There is also concern that the statistics used to calculate probability of pattern occurrence may not be representative of the occurence of DNA fragment patterns in the general population. A random distribution of genetic markers in the population requires a freely interbreeding and mobile population, which is not a general aspect of the human social condition. A DNA fingerprint pattern that is extremely rare in New York City, for example, can be extremely common in a small isolated town with a high degree of inter-family marriage. The sample populations used to establish the statistics regarding DNA fingerprint pattern frequency may not be large enough to allow application of the statistics outside of the original population. The frequency of occurrence of a DNA fingerprint pattern in New York City, for example, may be quite different from the frequency of occurrence of the same pattern in a small town in New Mexico.

Because the methods are scientically correct, the concerns regarding quality, reliability, and statistics can all be addressed and corrected. There is, however, a key component of the legal system that is more difficult to alter - the jury. In order to make a reliable decision, it is important that the members of the jury understand the capabilities and limitations of the method in use. For example, while a DNA fingerprint may conclusively prove two DNA samples are different, it **cannot** prove two samples identical. It merely provides a statistical argument about probability that two samples are identical. Two DNA samples can have the identical DNA fingerprint, yet have different nucleotide sequences (Figure 11.6). Resolving the controversy surrounding application of molecular biology methods to forensic and legal problems will require modifications within the fingerprinting companies and a reasonable general public understanding of molecular biology.

Generation of vaccines

The eradication of the variola, the virus that causes smallpox, is without question one of the greatest medical achievements of this century. The elimination of this disease was accomplished by innoculating virtually everyone in the world with vaccinia, a virus that causes very few symptoms in people but, because of similarity in the surface proteins of vaccinia and variola viruses, raises immunity against the smallpox virus. Since the variola smallpox virus can only propagate in a human host, as increasing numbers of people became immune to variola by vaccination with the related vaccinia virus, the ability of the smallpox virus to spread through the population decreased. Because of the lack of an infectable host, the variola virus eventually disappeared.

The strategy that allowed elimination of smallpox was possible because of a convenient similarity between the surface proteins of the disease-producing variola virus and the relatively harmless vaccinia virus. The similarity between the viruses that allowed the application of this

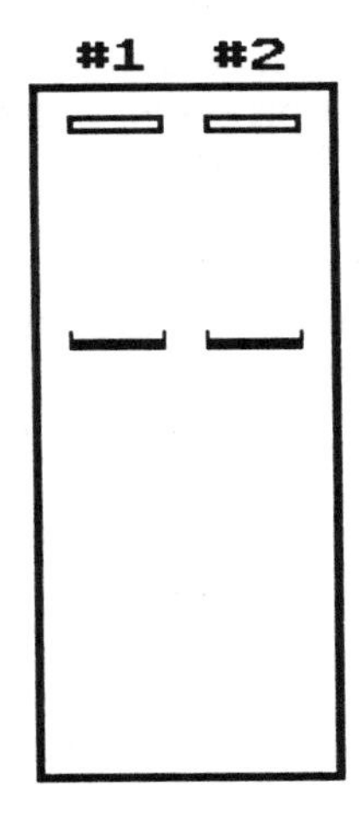

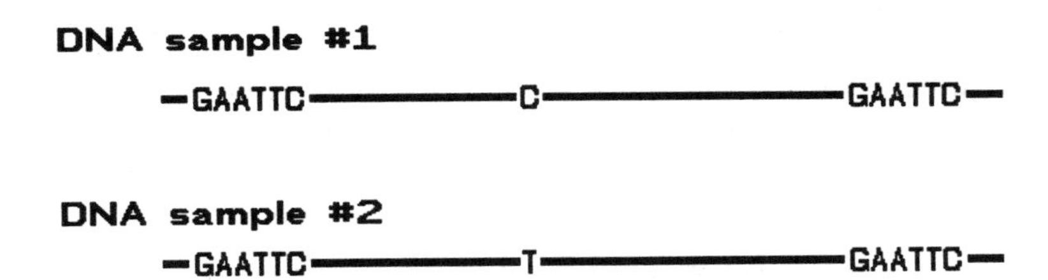

Figure 11.6. DNA bands of the same size in a DNA fingerprint are not necessarily identical. Two bands that appear to be identical on a DNA fingerprint, as illustrated, may actually have minor differences in nucleotide sequence. The difference of a single nucleotide residue would be sufficient to prove the although the two bands have the same size, they are different nucleotide sequences isolated from different pieces of DNA. It is relatively easy to obtain identical DNA fingerprints from two different, completely unrelated DNA samples.

strategy was coincidental and has not generally been found to occur naturally for other combinations of viruses.

Molecular biology has provided the methods that allow this strategy to be applied to generate immunity to other disease-producing viruses (Figure 11.7). The DNA for a harmless virus like vaccinia can be cleaved with a restriction enzyme and a gene that produces one of the surface proteins of a disease-producing virus inserted into the vaccinia virus. When the vaccinia construct is then used to vaccinate an animal, replication of the recombinant causes the production of not only vaccinia proteins, but also of the extra protein corresponding to the DNA insert. This will cause the formation of antibodies against the viral protein. If the immunized individual is subsequently exposed to the intact infectious virus, the antibodies directed against the coat protein that was carried by the recombinant virus can bind to the protein and inactivate the infectious virus.

The strategy with recombinant vaccines is identical to that used to raise immunity against the smallpox virus - infect the host with a harmless virus that causes the production of antibodies that recognize and inactivate a disease-producing virus. Molecular biology has now circumvented the need to find a naturally existing virus for vaccination. A virus like vaccinia can be used to carry genes isolated from other viruses and raise antibodies that recognize and inactivate the products of the cloned genes. This strategy has been used to generate effective vaccines against a variety of diseases, including rabies and hoof and mouth disease, and is under investigation as a means of generating vaccines against the AIDS family of viruses.

Manipulation of agronomically significant species

Molecular biology methods have been seen as having tremendous potential for the manipulation of the properties of agronomically significant species. The principal goal of this type of research is to use these methods to isolate genes for desired traits and introduce these genes into plant and

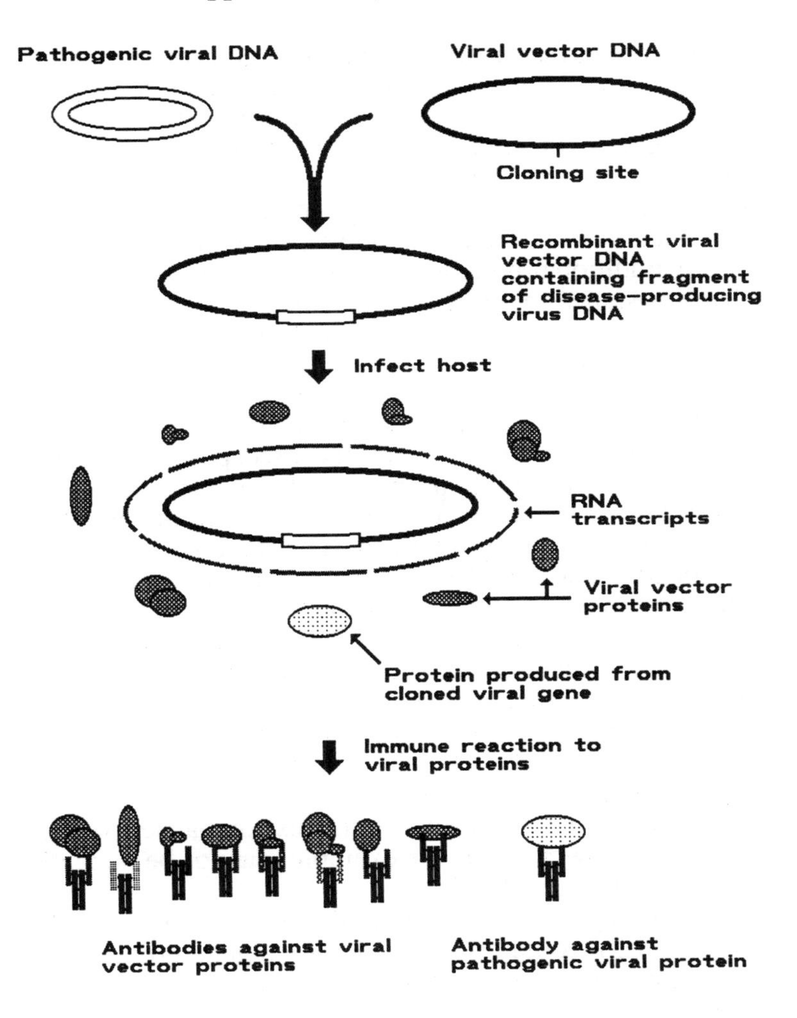

Figure 11.7. Viruses that cause only mild symptoms during infection of a host can be used as vectors to carry a gene from a pathogenic virus that normally causes a disease. When the recombinant virus is introduced into the host, replication of the recombinant virus will cause synthesis of viral RNA and proteins. RNA and protein will also be produced from the cloned gene but since the complete pathogenic virus is not present, no disease symptoms will be generated. The immune system can respond to the viral infection by making antibodies against both the normal viral proteins and the cloned viral gene product. The antibody against the pathogenic viral protein can subsequently confer immunity to infection by the intact pathogenic virus.

animal species that are used for the generation of agricultural products like food, fuels, and textiles. The concept of improving strains is not new. Man has for centuries been using selective breeding and genetic selection to genetically engineer plants and animals.
Molecular biology techniques merely facilitate the introduction and establishment of specific desired traits.

A traditional breeding scheme designed to introduce plant resistance to an insect pest might survey all related plants to find a "resistant" plant, a close plant relative that produces a noxious or poisonous compound that repels or kills insect pests. The resistant plant would then be genetically crossed with the crop plant and the progeny crop plants examined for resistance to the insect pest. Since genes that affect the desirable properties of the crop plant (such as yield or flavor) may be introduced during the breeding process, the resistant progeny must be examined relative to unwanted changes in desirable characteristics. Successive rounds of breeding with the crop plant may be necessary to re-establish desired traits in the resistant strain. With successive rounds of breeding and selection, the desired insect resistance trait might be bred into the plant to generate an improved crop plant (Figure 11.8). While this strategy has been successful for hundreds of genetic modifications of plants and animals, the process can be tedious, slow, and limited by sexual and species transmission barriers (genes from a sunflower plant, for example, cannot be sexually transferred to a potato plant).

A strategy that involves molecular biology methods (Figure 11.9) would also first identify a gene product that will provide the desired insect resistance, but the species containing the desired gene need not be present in a plant related to the crop of interest. The gene may be identified in a completely unrelated plant species, an animal, or a strain of bacteria. The desired gene would then be isolated, characterized, and fused to a transcription regulation region that will function in the crop plant. The engineered resistance gene can then be introduced by transformation into the desired plant host and progeny that express the desired trait selected. Unlike the convential genetic approach, the desired trait is the only gene introduced during

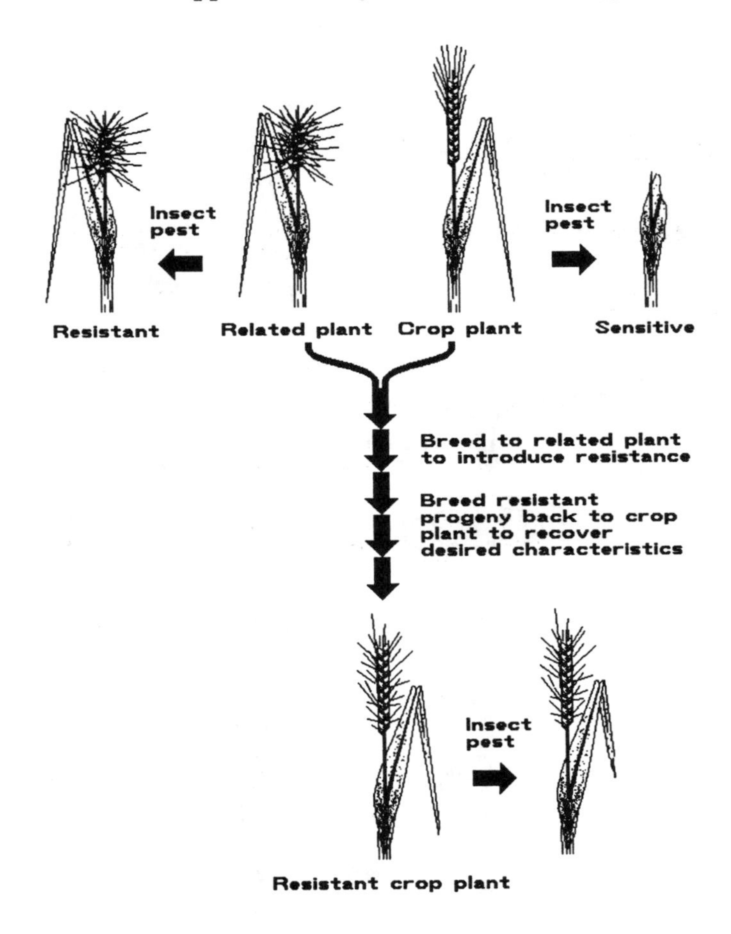

Figure 11.8. Introduction of a desired trait that is not normally present in an agronomically important species can be accomplished by conventional breeding methods. To introduce insect resistance in a crop plant, a breeder might screen other closely related plants to find an insectresistant plant that is sufficiently related to the crop plant to allow breeding of the two plants. The insect-sensitive crop might be crossed with the insect-resistant relative multiple times in order to acquire and stabilize the desired resistance trait. This breeding program may also introduce undesired characteristics of the insect-resistant relative, so insect-resistant progeny of the cross may be bred back to the crop plant parent to recover desired crop characteristics. Obtaining the desired insect-resistant crop plant may require many generations of breeding and may result in loss of some desireable crop plant characteristics.

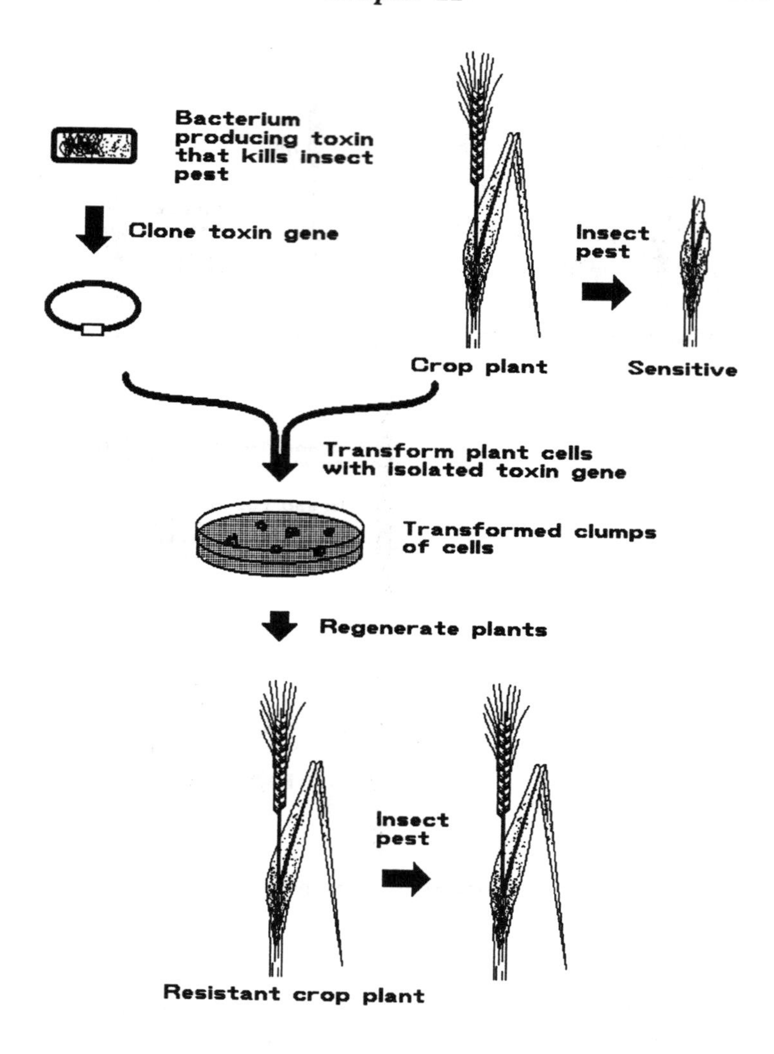

Figure 11.9. Introduction of desired traits can also be accomplished for increasing numbers of organisms by using molecular biology methods. Introduction of resistance to an insect pest might be accomplished by first identifying a bacterial strain producing a toxin that kills the insect pest. The gene producing this toxin could then be cloned in a plasmid vector and used to transform cells from the sensitive crop plant. Following isolation of transformed cells and regeneration of plants, the regenerates would be screened for an isolate with the desired insect resistance.

the procedure and repeated rounds of breeding are not necessary to remove undesired traits frequently acquired during conventional genetic schemes.

Molecular biology methods are currently being applied to a variety of agronomic problems to transfer genes across genetic species barriers and create plants and animals that express novel gene products. This application of these methods has led to public concern regarding a variety of issues. It has been suggested that genes introduced into a crop plant may subsequently be inadvertently sexually transferred to closely related weed species, that manipulation of the genome may lead to a decrease in genetic purity of a species, and release of the engineered strains into the environment may adversely affect native species. In a widely publicized controversy regarding the use of a strain of bacterium designed to outcompete resident bacteria involved in ice crystal nucleation and frost damage of strawberry plants, allegations were made that release of the genetically engineered bacteria might result in alteration of atmospheric weather patterns.

A wide variety of concerns have been raised regarding the use of genetically engineered organisms. These issues range from scientific concerns regarding the release of engineered organisms into an increasingly stressed biosphere to allegations regarding the morality of manipulating the genes of organisms. Perhaps because of the power of molecular biology to manipulate and modify the genomes of organisms, there has been significant emphasis on the potential contribution of this technology to agricultural production. As is often the case, the development and application of this powerful new technology has been accompanied by increasing amounts of public concern and controversy.

Summary

Molecular biology methods have tremendous value not only in the investigation of basic scientific questions, but also in application to a wide variety of problems affecting the overall human condition. Disease prevention and treatment, generation of new protein products, and manipulation of plants and animals for desired phenotypic traits are all applications that are routinely addressed by the application of molecular biology methods. Because of the wide applicability of these methods, they are rapidly becoming a pervasive - some would argue invasive - aspect of our technologically based society. The public concerns that address the application of these methods should be addressed by informed public discussion and debate. While scientists can be extremely critical of the quality, interpretation, and significance of experimental results, they have a rather remarkable tendency to be non-judgmental of the relative social merits of many applications of scientific research. It remains a public responsibility to be sufficiently well-informed to critically assess the merits of applied science research and participate in a communal decision-making process regarding the extent to which a new technology will be allowed to affect society.

12

Introductory Experiments in Recombinant DNA

"Few things are harder to put up with than the annoyance of a good example." — Mark Twain

The exercises that contained in this chapter have been chosen to demonstrate the basic principles in recombinant DNA: digestion of DNA with a restriction endonuclease, gel electrophoresis of DNA samples, insertion of DNA into cells can change their growth characteristics, and DNA can be rearranged to cause changes in genetic properties. The exercises are intended to be an example of the principles that DNA equals genes and that changes in DNA cause changes in genetic properties.

Each exercise is preceded by a short discussion of the principles involved and the specific goals of the procedure. Solutions that must be prepared for each exercise are detailed in a **Materials** section. While some of these "recipes" can be changed considerably without affecting the outcome of the exercise, many of the "minor" details are crucial to the success of the protocol. For example, a restriction digestion buffer may contain only 6 mM NaCl relative to 50 mM TRIS-Cl buffer, an apparently trivial amount of salt in the overall scheme of things. However, many enzymes are very specific about salt concentrations required for activity, and deleting the NaCl from the buffer may inactivate or actually change the recognition properties of the enzyme. Please do not arbitrarily change recipes and expect exercises to work properly.

The recipes assume a simple knowledge of chemistry and use standard abbreviations regarding concentrations:

M = moles/liter = molar
mM = millimoles/liter = 10^{-3} moles/liter = millimolar
l = liter
ml = milliliter = 10^{-3} liter
µl = microliter = 10^{-6} liter

When pH of a solution is indicated, there is generally an allowance of about 0.5 pH unit. If a pH of 7.5 is indicated, a pH of 7.0 to 8.0 will generally suffice. If possible, use a pH meter in the preparation of solutions, otherwise, use pH paper and be as accurate as reasonable.

The instructions often require the addition of water, sometimes indicated specifically as distilled water (dH_2O). For electrophoresis, media, and general solutions, tap water will often suffice. For reactions involving enzymes, distilled water should always be used rather than tap water. Many enzymes can be inhibited by heavy metal salts present at low levels in tap water. Distilled water purchased for use in a steam iron can be used in the event that there is no access to a water still or deionizer.

Certain solutions must be sterilized (media for culture of bacteria, for example). An electric hot plate and a standard pressure cooker can be used as a substitute for an autoclave. Heat media at 16 lb pressure for 20-30 minutes to sterilize small volumes of liquids (less than 500 ml solution per container). While a microwave oven is used for melting agarose in nearly every molecular biology research lab, a hot plate with a boiling water bath or a gas burner will accomplish the same thing. Agarose solutions have a tendency to superheat and boil violently when the agarose granules are first melting. A solution of agarose that has been previously melted, then allowed to cool and solidify is less likely to boil violently when melted the second time.

As a note of caution, realize that molecular biology uses a number of noxious chemicals: organic solvents (phenol, chloroform, ether) are quite toxic and certain compounds (ethidium bromide and UV light, for example) are known mutagens capable of causing genetic changes. The use of such compounds has been kept to a minimum in these exercises and where such compounds are used, precautions are noted in directions for the exercises. During electrophoresis, although low voltages are used, severe injury is possible. Electrophoresis boxes should have an interlock mechanism - a device that prevents the current from being applied to the buffer when the buffer is exposed. Most commercial gel boxes include this safety design feature, as do the plans that accompany these exercises. COMMON SENSE IS REQUIRED IN ALL LAB EXERCISES. Food and drinks should be prohibited from the work area and lab coats should be worn. Power units should be turned off while loading gels or handling gel boxes. All spills should be cleaned up when they occur.

Having previously indicated that these exercises should not be arbitrarily altered, I now emphasize that these protocols should not be considered inviolate. There are about 113 ways to accomplish the same thing in any molecular biology exercise. As your understanding of these methods increases, you will recognize how protocols can be altered to suit a particular circumstance. Other protocols will appear that seem easier or more reproducible in your own situation. While this is one of the aspects of molecular biology that makes the methods so powerful, this variability also tends to unnecessarily confuse novices in the field. Before modifying a protocol, be certain that you understand the changes and that these alterations will not adversely affect the outcome of an exercise.

The bacteriophage lambda, pBR322, and pUC19 plasmid DNA samples used in these exercises are available from a number of commercial suppliers. The bacterial DNA can be prepared by a simple procedure (see Appendix) that uses detergent lysis and phenol/chloroform extraction to prepare high quality DNA. Commercially available calf thymus DNA can be substituted. Plasmid recombinants containing *E. coli* DNA in pUC19 were chosen specifically to work entirely with bacterial DNA and minimize any misperception of biohazard potential. Competent cells are commercially available from several sources, but acceptable competent cells can also be prepared with a minimum of expertise (see Appendix).

Exercise 1. Gel electrophoresis of nucleic acids

Nucleic acid samples are often subjected to gel electrophoresis to characterize the size and number of different fragments in the sample. In this exercise, DNA samples that have been digested with restriction enzymes and mixed with a tracking dye will be subjected to agarose gel electrophoresis. The electrophoresis buffer, the salt solution that both conducts electric current and controls the pH of the solution during the separation of the DNA fragments, is TRIS/borate/EDTA or TBE, a commonly used buffer system.

When charged molecules are placed in an electric field, the molecules will migrate towards one of the electrodes, depending on the net charge of the molecule. Nucleic acids have an overall negative charge due to the negative charges associated with the phosphate backbone of the molecules, so they will migrate towards the positive electrode. Since the distribution of phosphate is very regular across the length of the nucleic acid molecule, nucleic acids have a constant charge/mass ratio and will therefore migrate at the same rate in an electric field.

Separation of nucleic acid molecules of different size or conformation is achieved by adding a support matrix to the electric field and forcing the molecules to migrate through this matrix, typically agarose or polyacrylamide gel. This matrix acts like a sieve that allows small molecules to go faster than large molecules. The result is that molecules separate in the matrix according to their relative size and shape.

This exercise will use gel electrophoresis to examine the fragments present in several DNA samples. The principle goal is to obtain experience in pouring gels, loading DNA samples, and visualizing the DNA bands.

Due to the small volumes of sample to be applied to the gel, samples are routinely handled with a manual micropipet device that uses a disposable plastic tip to handle a 10 to 20 µl sample. Several commercial devices are available. A substitute sample loader can be assembled from a disposable 0.5 or 1 ml syringe fitted with a disposable plastic tip (see Appendix).

Materials

- *Eco*RI- and *Hin*dIII-digested bacteriophage lambda DNA samples containing SM Dye (other DNA samples can be added or substituted).

- Reaction Stop Mix Dye (SM Dye): 10% glycerol, 0.5% SDS (sodium dodecyl sulphate), 0.025% xylene cyanol FF dye (XC), 0.025% bromphenol-blue WS dye (BPB). This mix is added to a DNA sample after digestion to prepare the sample for electrophoresis. The SDS helps inactivate DNA binding proteins and releases them from the DNA fragments and the glycerol weights the sample so that it will layer uniformly into the slots in the gel. The two dyes serve as visual markers of the progress of the electrophoresis only and do not stain DNA or proteins. The purple BPB dye will migrate with about twice the relative mobility of the turquoise XC dye.

- TBE electrophoresis buffer: 90 mM TRIS, 2.5 mM Na_2EDTA, 89 mM boric acid, final pH 8.2 (can be stored as a 10x stock).

- 1% Agarose: Molten 1% agarose in TBE buffer, heated to 100°C to melt the agarose and stored in 55°C water bath until needed. Melt the agarose very carefully to minimize superheating and violoent boiling.

- Ethidium bromide (EtBr): For staining gels to visualize DNA, approximately 200 ml of 4 µg/ml ethidium bromide in water in a shallow tray, stored covered and protected from excessive light exposure. EtBr should be treated as a mutagen - a compound capable of causing genetic mutations - and bare skin should not come in contact with the solution. It is both photosensitive and biodegradable and **dilute** solutions are often disposed of via the sink. **Concentrated** solutions must be saved and disposed of as biohazardous chemical waste.

Protocol

1. Use the molten agarose to pour a minigel as demonstrated or according to instructions for the gel unit in use. Illustrations for assembling a generic gel unit are included in the Appendix. It is important to be certain that all small agarose particles are completely melted before use. Allow the gel to completely harden before removing the well-forming comb. This should take about 15 minutes.

2. Remove the comb from the solidified gel and transfer the gel to a running unit. Submerge the gel in TBE buffer.

3. Use a manual pipettor with a disposable plastic tip to load 10 µl of each sample into one of the wells in the gel. Plasmid and bacteriophage DNA samples should be easy to load, but chromosomal DNA samples may be quite viscous and difficult to load.

4. Once the samples have been loaded, close the unit and apply power to the electrophoresis chamber. Typical minigels run at 100-130 volts (60-160 milliamps), depending on the unit used. During the run, the XC and BPB dyes will resolve into two bands, with the BPB the faster band. Run the gel until this band is near the end of the gel, then turn the current off and remove the gel.

5. Transfer the gel to the staining tray containing the ethidium bromide solution, and stain the gel for 5 minutes. Ethidium bromide is a mutagen, and gloves should be worn or a spatula used to transfer the gel in and out of the ethidium solution.

6. Transfer the stained gel to a destaining tray containing water. Destain the gel for 5 minutes to remove excess ethidium bromide.

7. The stained gel can be visualized with a UV transilluminator or a UV mineral lamp. DO NOT LOOK DIRECTLY AT THE UV LIGHT!! UV light causes skin

burns, is a mutagen, and will cause severe headaches from eye damage with direct exposure. Always use a face mask or shield.

Analysis and significance of results

The results of a typical gel are shown illustrated below:

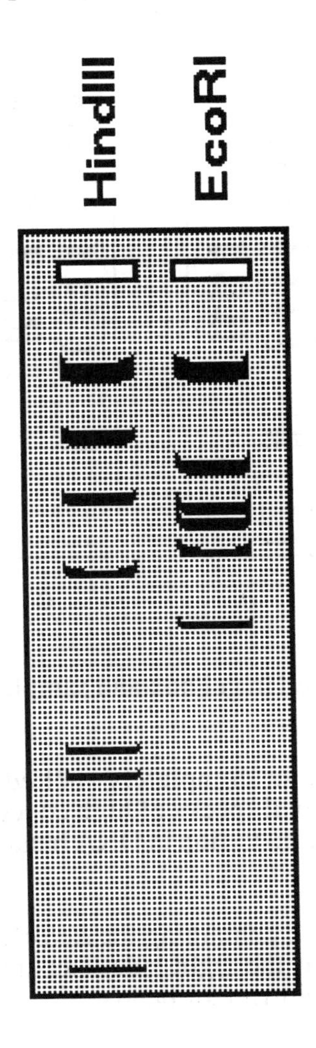

If lambda DNA samples digested with different restriction endonucleases are present on the same gel, note that each different restriction enzyme produces a distinct, completely reproducible pattern of DNA fragments. Relative separation of individual fragments is dependent on agarose concentration, electrophoresis conditions, and the quality of the gel preparation.

Many different types of artefact can distort the bands in a gel. Some of the more common gel artefacts are illustrated below.

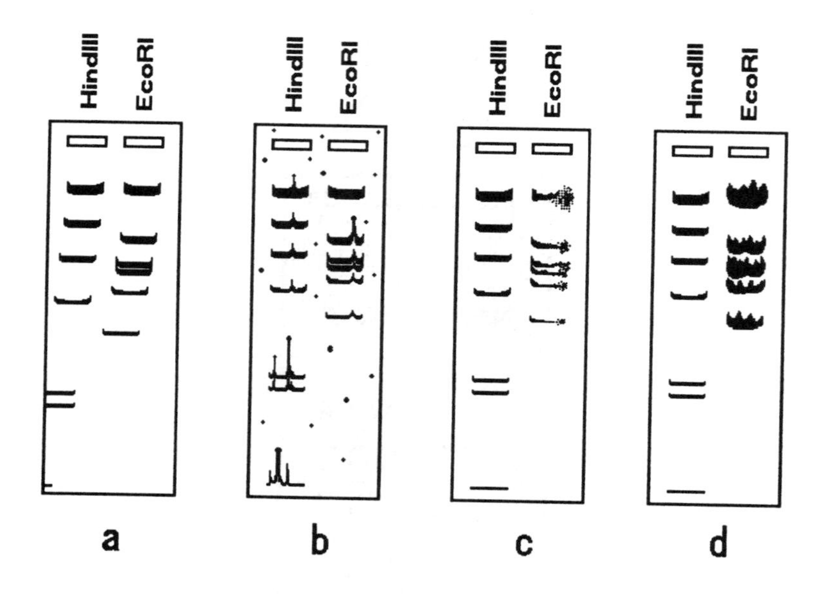

a. Gel was crooked in gel box and samples migrated off of the gel.

b. Agarose was not completely melted before gel was poured. The small specks of high concentration agarose in the gel cause distortion in DNA bands and bright spots in the gel.

c. The sample well containing the *Eco*RI-digested lambda DNA leaked at the right side, causing loss of sample intensity and blurring of bands.

d. The smearing of the *Eco*RI-digested lambda DNA is caused by loading too much DNA.

During electrophoresis of DNA fragments through an agarose gel, the fragments separate by size with smaller fragments moving faster than larger fragments. The relationship between size and mobility is not linear, but can be approximated by plotting the distance migrated versus the log(Molecular Weight).

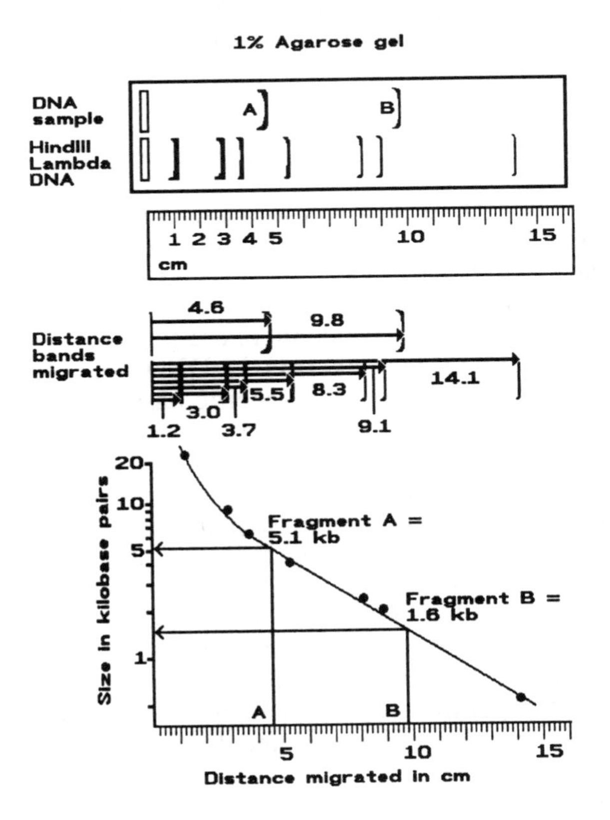

Using the *Hin*dIII fragments as molecular weights standards to estimate the size of other DNA fragments. Mobility of fragments is determined and plotted as log(Size) versus mobility. The line drawn through the lambda DNA fragments can be used as a standard curve for determination of sizes of other DNA fragments present on the same gel.

Exercise 2. Restriction endonuclease digestion of chromosomal and plasmid DNA

Digestion of DNA with a site-specific restriction endonuclease will generate a set of specific DNA fragments. In this exercise, a variety of DNA samples will be digested with the restriction enzyme *Eco*RI. This enzyme recognizes and cleaves the sequence 5'-GAATTC-3'between the G and A residues to generate the four-base cohesive terminus AATT:

```
5'-NNNNGAATTCNNNN-3'   EcoRI    5'-NNNNG        AATTCNNNN-3'
3'-NNNNCTTAAGNNNN-5'  ------>   3'-NNNNCTTAA        GNNNN-5'
```

Because *Eco*RI recognizes and cleaves at a six-base sequence (5'-GAATTC-3'), the number of fragments generated by digestion of a DNA molecule with this enzyme is determined approximately by the frequency of occurence of the recognition site. Since there are four possible bases that can occur at any position in DNA (A,C,G,T) the probability of occurence of a specific base is (1/4), while that of a specific two-base sequence is $(1/4)^2$, a four-base sequence $(1/4)^4$, and a six-base sequence is $(1/4)^6$. The six-base *Eco*RI site, therefore, should occur about once every 4^6 bases, or once every 4096 base pairs. *Eco*RI cleaves the 2,800 base pair plasmid pUC19 only once and will convert the circular molecule to a linear form, cleaves the linear 50,000 base pair bacteriophage lambda DNA five times to generate six fragments, cleaves hundreds of times in the *E. coli* DNA genome, and thousands of times in eukaryotic DNA. The larger the genome of an organism is, the greater the number of fragments generated during digestion with a restriction enzyme.

Several DNA samples from genomes of increasing size will be digested with *Eco*RI in this exercise. If both the enzyme digestion and the gel analysis are to be performed in the same day, as soon as the digestion reactions have been set up and are incubating, prepare agarose minigels for analysis of the digested DNA samples.

Materials

- Plasmid pUC19 vector DNA: 100 μg/ml.
- Bacteriophage lambda DNA: 100 μg/ml.
- Bacterial chromosomal DNA: 100 μg/ml.
- *Eco*RI endonuclease: 3-5 units/μl. **Keep this in ice at all times!!**
- 10x *Eco*RI Reaction buffer: 500 mM TRIS-HCl pH 7.0, 750 mM NaCl, 60 mM $MgCl_2$, 60 mM 2-mercaptoethanol. Most enzyme companies supply the correct buffer with each enzyme purchased.
- Water (dH_2O): Sterile, deionized or distilled.
- Manual pipet devices, 1-20 μl and 20-200 μl capability.
- Disposable conical 1.5 ml microfuge tubes and pipet tips.

Protocol

1. For each different chromosomal DNA sample, set up a digestion reaction in a labeled 1.5 ml conical tube by making the following additions:

10x *Eco*RI buffer	5 μl
dH_2O	35 μl
1 ug DNA	10 μl
Total volume	50 μl

Flow chart, Exercise 2

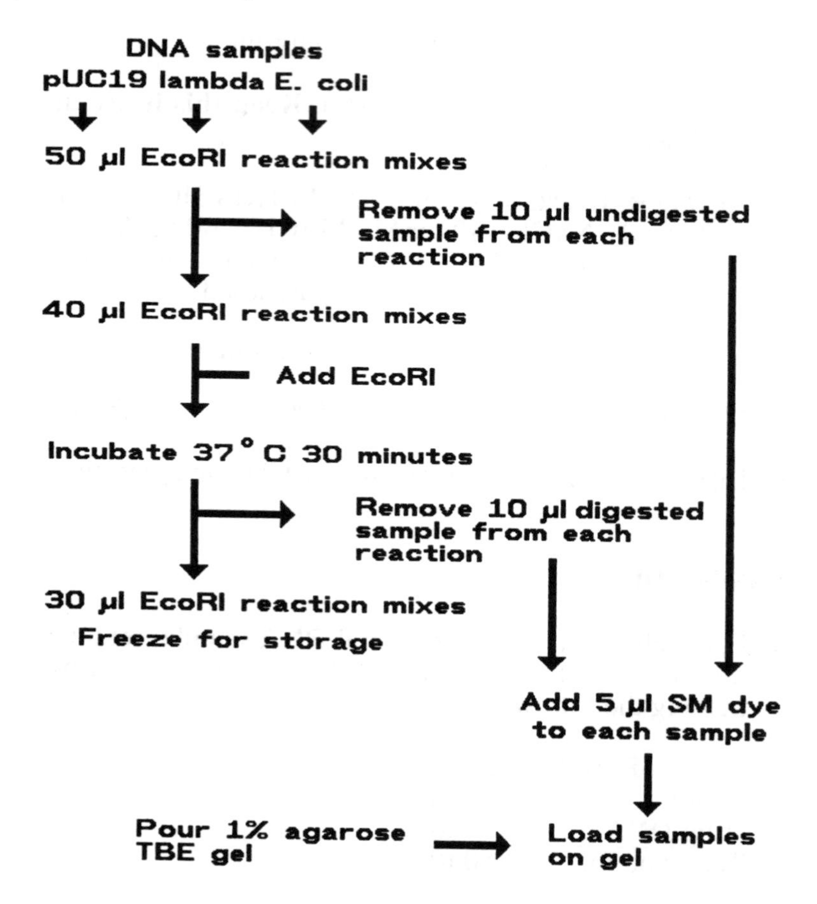

2. Mix the contents of the tubes gently but thoroughly, then from each tube remove 10 μl and place in a tube labeled with the sample number and the term "-EcoRI". Place these 10 μl samples on ice. They are the undigested controls and will be used to compare to samples following digestion.

3. To the remaining 40 μl of each sample add:

   *Eco*RI, 5-10 units   2 μl

   Mix, then incubate at 37°C for 30 minutes.

4. From each digestion tube, remove 10 μl of each sample and add to a tube labeled with the sample name and "+EcoRI". The remainder of each digestion reaction should be frozen for further use.

5. You should now have two tubes containing 10 μl of each DNA sample, one "-EcoRI" and one "+EcoRI", for a total of six tubes. To each of these tubes add:

   SM dye   5 μl

   Mix the contents of each tube, and the samples are ready for electrophoresis. Samples can be stored on ice or frozen for future gel electrophoresis.

Significance of procedure

This exercise has prepared reaction mixes that contain several DNA samples of different sizes and conformations, then digested portions of each reaction mix with the restriction enzyme *Eco*RI. The samples are now ready for analysis by agarose gel electrophoresis to determine the effects of the restriction enzyme digestion. The same gel electrophoresis described in Exercise 1 will be utilized.

Materials

- DNA samples prepared in Exercise 2.
- Electrophoresis materials, see Exercise 1.

Protocol

1. Prepare a 1% agarose gel in TBE buffer as in Exercise 1.

2. When the gel has solidified, assemble the gel unit. Load 15 μl of each of the six samples. You may also want to include one lane of the digested lambda DNA used in the first exercise as a DNA standard. Electrophorese as in Exercise 1.

3. Stain the gel with ethidium bromide and visualize the DNA bands with a UV light box. A photograph of the gel is useful, but a sketch of the gel is sufficient for recording information.

Analysis and significance of results

A typical gel is illustrated below:

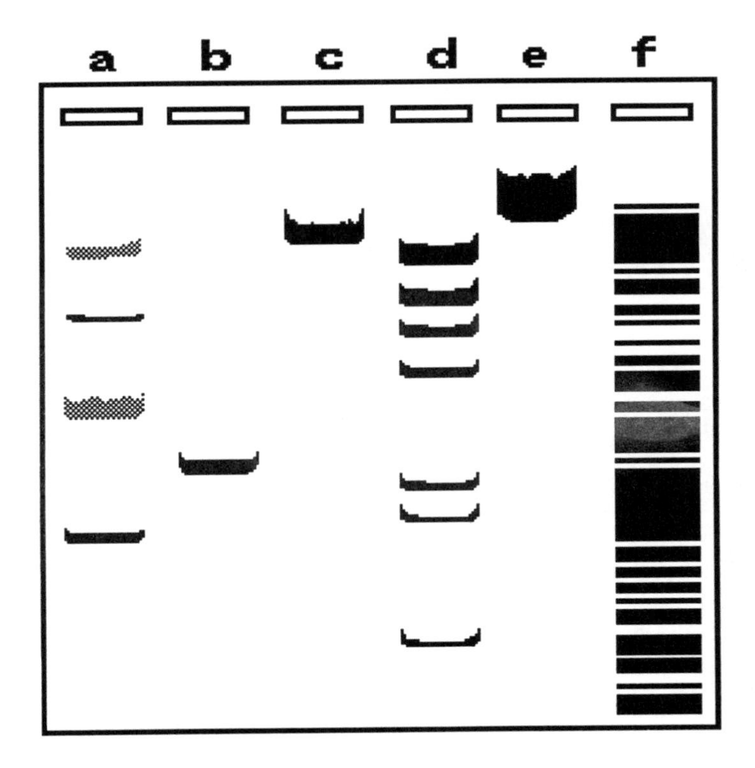

Compare each undigested (-*Eco*RI) with the corresponding digested (+*Eco*RI) sample. Note that the migration position of the undigested circular plasmid pUC19 DNA (a) changes when cleaved at the unique *Eco*RI site to generate a linear molecule (b). The linear undigested lambda DNA (c) contains five *Eco*RI cleavage sites, and migrates as six bands after digestion (two of the bands are of similar size and can be difficult to resolve, but this double band will be twice as intense as the other bands) (d). The slowly migrating, broad, undigested bacterial chromosomal DNA sample (e) is converted upon digestion to hundreds of smaller, yet distinct bands (f). Note that as the size of the genome (pUC19<lambda<*E. coli*) increases, the number of cleavage sites also increases, demonstrating that larger genomes are actually composed of larger amounts of DNA.

Exercise 3. Competent cell production

Although some bacteria naturally undergo a growth phase when they are capable of taking up DNA, strains of *E. coli*, the bacteria most commonly used in a molecular biology lab, cannot normally take up DNA. These bacteria must be chemically treated to be made competent for transformation. Competent cells can be purchased from a variety of commercial sources but can be relatively expensive for routine classroom demonstrations and laboratories. Cells can be made by harvesting rapidly growing bacteria and exposing them to calcium chloride on ice. The following protocol can be used to make competent *E. coli* that are of sufficient quality for routine use.

Materials

- 35 ml LB medium (1% Bacto Tryptone, 0.5% Bacto Yeast Extract, 0.5% NaCl, pH 7.0-7.2) in sterile 100-250 ml flask.
- Fresh 2 ml overnight LB culture of JM83.
- Sterile 50 or 35 ml centrifuge tube.
- Sterile 30 mM $CaCl_2$.
- Sterile 30 mM $CaCl_2$, 15% glycerol.
- Sterile microfuge tubes.
- Ice.
- Refrigerated centrifuge. If a refrigerated centrifuge is not available, place a clinical or counter-top centrifuge in a refrigerator. Allowing the cells to warm up during the centrifugation may affect the competency of the cells.

Protocol

1. Competent cells must be made from a rapidly growing culture of bacteria to obtain good efficiencies of transformation. Mid-logarithmic phase of growth (3-5 x 10^8 cells/ml) is good for preparation of general-use competent cells. With a fresh overnight innoculation culture, a 1:50 dilution into a fresh culture will generally give mid-log cells in 60-90 minutes of growth at 37°C. If an old overnight culture is used or if insufficient dilution is made, the bacteria will undergo a significant lag phase before growth begins.

2. Use sterile techniques to transfer 0.7 ml of the overnight culture of JM83 to the flask containing 35 ml of LB. Incubate in a 37°C shaker with vigorous shaking to allow growth. Monitor the growth of the culture every 20 minutes (A_{550}). When absorbance is approximately 0.5, the culture is ready to harvest. This should correspond to a mid-logarithmic phase of growth. If a semi-log plot of A_{550} versus time is prepared, harvest of cells should occur in the linear portion of the graph.

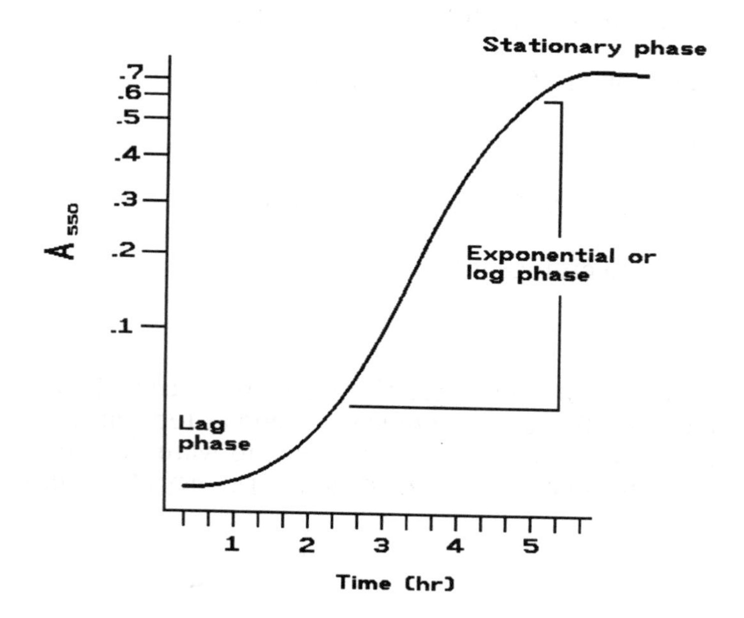

3. Although strict sterile technique is not necessary, try to minimize contamination of the cells throughout the following procedures. Pour the culture into a sterile centrifuge tube. Spin 7000 rpm, 5 minutes to pellet the cells.

4. Pour off the supernatant. Add 17.5 ml of cold 30 mM $CaCl_2$ (1/2 of the original culture volume). Gently resuspend the cells.

5. Place cells on ice for 20 minutes. Cells will swell to form spheroplasts. Cells can incubate on ice as long as 90 minutes.

6. Centrifuge 7000 rpm, 5 minutes, at 4°C.

7. Pour off supernatant. Many strains of *E. coli* often form a halo instead of a normal pellet.

8. Add 3.5 ml of cold 30 mM $CaCl_2$, 15% glycerol. GENTLY RESUSPEND THE CELLS. If resuspension is too harsh, competency may decrease. Return cell suspension to ice.

9. Label 17 sterile microcentrifuge tubes "cJM83". Aliquot 200 µl of competent cells into each tube. Cap tubes. Cells can be used immediately or stored in a freezer for future use. To freeze, move the tubes to a pre-chilled rack in the freezer for storage. Cells can be stored at -20°C (normal lab freezer) for several weeks or at -70°C for months without significant loss of competency. Recombination deficient strains (*rec*A$^-$ strains, in particular) rapidly lose viability/competency when stored in a freezer.

10. If competent cells are made and stored for use in a lab exercise, it is wise to transform one tube with a pure plasmid DNA sample prior to use to verify that the cells have not lost competency during storage. Larger numbers of competent cells can be prepared by scaling all procedures up as needed. When culture volumes are

increased, be certain that the culture has a large surface area to maintain adequate aeration in the growing culture.

Exercise 4. Transformation of competent cells with pUC19 and pUC19 recombinants

DNA molecules can be inserted into bacteria by first chemically treating the bacteria to make them competent, or able to take up DNA. This chemical treatment is often accomplished by exposing the bacteria to specific salts that cause the bacteria to swell and become capable of binding DNA. Treatment of *E. coli* with $CaCl_2$ gives reproducible transformation results, although higher levels of competency can be attained with other procedures. After exposing the competent cells to DNA on ice, a 43°C heat shock is required to make the competent cells take up the bound DNA.

In this exercise, *E. coli* strain JM83 will be transformed with the plasmid pUC19. This plasmid contains genes that allow its own replication in the bacterium and contains a gene (*bla*) that encodes ß-lactamase, an enzyme that confers resistance to ampicillin. The host strain JM83 is sensitive to and will not grow in the presence of ampicillin, while JM83 containing pUC19, indicated by JM83(pUC19), will grow and form colonies in the presence of the antibiotic.

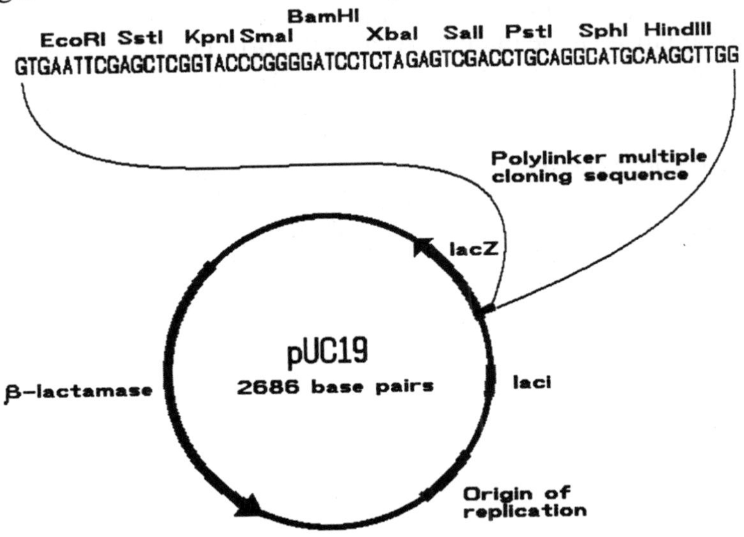

This plasmid also contains a genetically engineered portion of the *lac* operon of *E. coli*. When exposed to the inducing compound IPTG, cells containing pUC19 synthesize the enzyme ß-galactosidase. This activity can convert the colorless compound X-gal to a blue compound during the growth of the cells. When the strain JM83(pUC19) is plated on medium containing ampicillin, IPTG and X-gal, the resulting colonies will be dark blue in color.

The plasmid pUC19 was designed with a series of unique restriction cleavage sites within the coding region of the ß-galactosidase gene. This region, designated the multiple cloning site or MCS, is used for the specific insertion or cloning of extra DNA fragments. When a fragment is inserted, the β-galactosidase gene is interrupted and the enzyme can no longer be synthesized. Thus, a strain containing pUC19 with an extra DNA insert, designated JM83(pUC19:insert), will still form colonies in the presence of ampicillin, but in the presence of IPTG and X-gal the colonies will remain white rather than blue. Thus, bacteria containing the wild-type pUC19 vector can be distinguished from cells containing a recombinant pUC19 plasmid with an extra DNA insert.

To transform cells with pUC19, the competent cells will be placed at 43°C, then directly plated on nutrient plates containing 100 ug ampicillin/ml as a selective antibiotic. Incubation of the plates at 37°C will allow colonies to form. The presence of the compounds IPTG and X-gal will be used to distinguish the original vector from recombinant derivatives containing DNA inserts. The wild-type vector will cause the formation of blue colonies and recombinant derivatives present in the pUC19:*E. coli* DNA library will cause the formation of white colonies.

Flow chart, Exercise 4

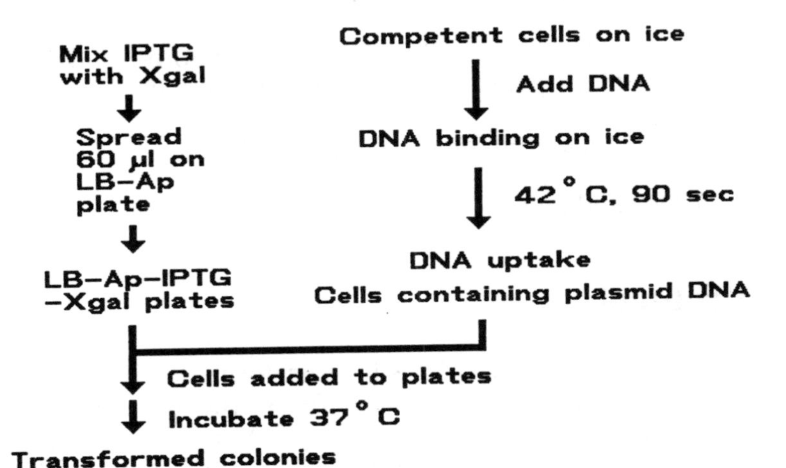

Materials

- Plasmid pUC19 vector DNA: 100 µg/ml.
- pUC19:*E. coli* library DNA: 100 µg/ml.
- Competent *E. coli* cells: $CaCl_2$ treated competent JM83 cells, 100 µl aliquots, stored -20°C (see Appendix).
- Sterile dH_2O.
- LB agar plates containing ampicillin for selection of pUC19 transformants: 1% Bacto Tryptone, 0.5% Bacto Yeast Extract, 0.5% NaCl, 1.5% Bacto agar, pH 7.0-7.2. Autoclave to sterilize, cool to 50°C, add ampicillin to 100 µg/ml, pour petri plates, and cool. Other complete *E. coli* nutrient media can be substituted, as long as ampicillin is present. Do not use minimal media, as most laboratory strains of bacteria have several nutritional deficiencies that require specific supplements that are generally present in rich media.

- 10 mM IPTG: 23.8 mg IPTG (isopropyl-β-D-thiogalactopyranoside)/ml dH_2O.
- 5% X-gal: 50 mg X-Gal (5-bromo-4-chloro-3-indolyl-β-D- galactopyranoside)/ml dimethylformamide.
- 43°C temperature block or water bath.
- 37°C incubator.

Protocol

1. Prepare six LB-Ap plates for transformation by adding IPTG and X-gal to each plate. To a sterile tube add 60 µl 10 mM IPTG and 300 µl 5% X-gal. Mix solution. Place 60 µl of this IPTG/X-gal mix on the surface of an LB-Ap plate. Dip a glass spreader in 95% ethanol and flame to sterilize. Use the spreader to distribute the IPTG/X-gal mix evenly over the surface of the plate. Replace the lid on the plate and invert the plate. Repeat this process for each of the remaining 5 plates. If you attempt do do all of the plates at the same time, the solution may soak into the center of the plates and will not be uniformly distributed over the surface. Allow the solution to soak into the surface of the plates for 15-30 minutes (if possible, place the plates in a 37°C incubator during this time period).

2. Thaw three tubes of competent cells at room temperature. As soon as cells are thawed, label one tube "-DNA", one tube "+pUC19", and one tube "+library" and place all tubes on ice.

3. To the "-DNA" tube add 10 µl sterile ddH_2O, to the "+pUC19" tube add 10 µl (1 µg) of pUC19 DNA, and to the "+library" tube add 10 µl of pUC19:EC library DNA. Place tubes on ice for at least 15 minutes (length of time is not crucial here). This allows DNA to bind to the competent cells. While tubes are on ice, adjust a water bath to 42-43°C.

4. Heat pulse the cells to cause DNA uptake. Transfer all three of the tubes of competent cells containing DNA to the 42°C bath for 90 seconds, then transfer to a room temperature rack.

5. Add 10 μl of one of the transformed cell samples to an LB-Ap-IPTG-Xgal plate and the rest (100 μl) to another plate. Dip the glass spreader in ethanol, flame, and use to distribute the cells evenly over the surface of each plate. Invert the plates and label with amount of cells and type of DNA (such as 10 μl, pUC19). Repeat for each of the tubes of transformed cells (you should end up with no tubes of transformed cells left over and have six labeled plates with cells spread on them). Place plates in a 37°C incubator for 12-18 hours to allow colonies to form and color to develop. Plates can be stored in a refrigerator and color will continue to develop.

Analysis and significance of results

After the plates have been allowed to incubate and colonies have formed and colored, compare all the plates. The following results are typical:

- JM83 transformed with no DNA should give no distinct colonies but may have some smearing or poor growth in areas of dense numbers of cells.

- JM83 transformed with pUC19 DNA should give many distinct colonies, nearly all of which (>99%) should be blue. The occasional white colony will be a contaminant or a mutant derivative of pUC19 in which the ß-galactosidase gene is no longer active. By counting the number of colonies on a plate and dividing by the amount of transforming DNA represented on the plate, it possible to calculate the efficiency of transformation in terms of transformants/μg DNA. Example: 1 μg of pUC19 DNA was added to 100 μl of competent cells, and 1 μl of this mix was plated on an LB-Ap plate, resulting in 227 colonies. The amount of DNA represented on the plate is 10 µl/110 µl x 1 µg = 0.1 µg. Transformation efficiency is 227 colonies/0.1 µg, or 2.27×10^3 transformants/µg DNA.

- JM83 transformed with pUC19:*E. coli* library should give similar numbers of colonies when compared with the pUC19 transformation, but only a percentage of the colonies will be blue. The blue colonies are the result of transformation with pUC19 containing no additional DNA insert, and the white colonies are caused by transformation with recombinant pUC19 plasmids. The transformation efficiency can be calculated for the DNA as described above and the percentage of recombinant molecules can be determined by the calculation:
  100 x # white colonies/# total colonies.

This exercise demonstrates two fundamental principles of recombinant DNA methods:

1. DNA can be inserted into bacteria to change the properties of the cells (ampicillin-sensitive cells converted to ampicillin-resistant colonies).

2. DNA fragments can be re-arranged to change the genetic properties of the DNA molecules (blue colonies caused by pUC19 are converted to white colonies when DNA fragments are inserted into pUC19 DNA).

These principles can be further examined by isolation and examination of plasmid DNA present in the transformant colonies.

Exercise 5. Transformation of competent cells with pUC19 and pBR322 plasmid DNA

One of the most important concepts of DNA biology is that genes are composed of DNA and that the introduction of DNA into a cell can change the physical properties of the cell. This can be illustrated with petri plates containing a simple nutrient medium (Luria broth, Tryptone, Nutrient agar, essentially any medium that will allow the growth of *E. coli*), a small amount of plasmid DNA, and competent *E. coli* bacterial cells.

This exercise uses *E. coli* JM83 cells that have been made competent, or able to take up DNA. The cells were grown to mid-logarithmic phase, then harvested and resuspended in 30 mM $CaCl_2$. After incubation on ice for 30 minutes, the cells were again harvested and resuspended in 30 mM $CaCl_2$ with 15% glycerol. Cells were dispensed into 0.2 ml volumes in sterile 0.5 ml tubes. Each of these tubes contains sufficient cells for at least 4 different transformation reactions. Competent cells may slowly lose their ability to take up DNA if not stored at -70°C. For this reason, the competent cells are shipped in dry ice to arrive the week before the exercise and are stored in a freezer at about -5°C until used.

A DNA sample containing two types of pure plasmid DNA, pUC19 and pBR322, at a concentration of 100 µg/ml will be used to demonstrate that DNA can change cellular properties. The plasmid pUC19, a common plasmid cloning vector used to carry and propagate other DNA fragments in bacteria, has two genes of interest: one confers resistance to the antibiotic ampicillin, and the other encodes a ß-galactosidase gene derivative that produces a protein that can help the bacteria metabolize derivatives of the sugar galactose. When the colorless compound X-gal is added to the growth medium, the ß-galactosidase protein can metabolize the X-gal into a colored derivative that turns the bacterial colonies blue.

The second plasmid DNA provided is pBR322, one of the older plasmid cloning vectors. Like pUC19, this plasmid confers resistance to ampicillin, but it also confers resistance to the protein synthesis inhibitor tetracycline. Just like bacteria containing pUC19, bacteria containing pBR322 will be able to grow in medium containing ampicillin. Since pBR322 does not contain the ß-galactosidase gene derivative, cells containing this plasmid will be unable to metabolize X-gal to a colored compound and will remain white when exposed to X-gal. Bacteria containing pBR322 will, however, be able to grow in the presence of a concentration of tetracycline that kills cells containing pUC19.

The bacterial strain JM83 is sensitive to ampicillin and will not grow on nutrient plates in the presence of the antibiotic at a concentration of 100 µg/ml. When either pUC19 or pBR322 is inserted into the competent JM83 cells, the cells become transformed and are able to grow in the presence of the antibiotic. These two plasmids can thus cause the same change in bacterial phenotype - ability to grow in the presence of ampicillin - and this property cannot be used to distinguish between cells containing one of these two plasmids.

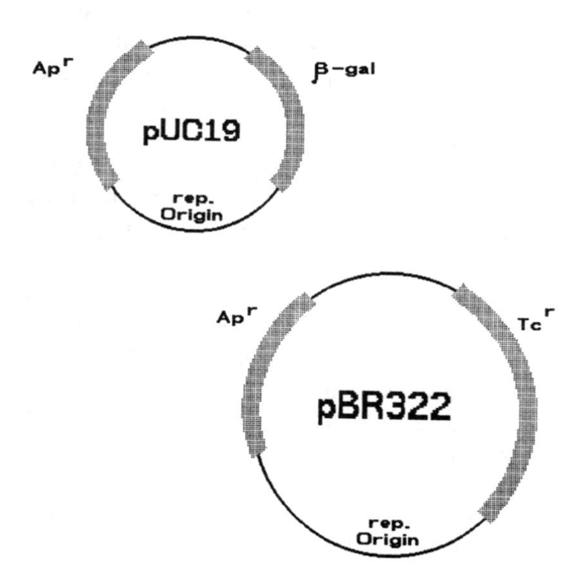

Genetic maps of the plasmids pUC19 and pBR322.

However, if the compounds X-gal and IPTG are also added to the ampicillin plate, the ß-galactosidase gene present on pUC19 will cause bacteria containing this plasmid to form blue colonies, while bacteria containing pBR322 will form white colonies. The difference in phenotype associated with these plasmids - ability of cells containing pUC19 to form a blue colony - allows rapid sorting of colonies formed by cells containing each of the two types of plasmid. This demonstrates that two plasmids that confer the same phenotypic property (such as ability to grow in the presence of ampicillin) can be readily distiguished if the plasmids differ in ability to confer a second phenotype (such as ability to metabolize X-gal to a blue compound).

This exercise will also demonstrate a second important principle of DNA manipulation: when two genes are both present on the same DNA fragment, selection for one of the genes will force maintenance of the second gene as well. As the transformed bacteria grow and form colonies in the presence of ampicillin but in the absence of tetracycline, there is no selective pressure that requires the expression of the tetracycline resistance gene present on pBR322. Nevertheless, because the tetracycline resistance gene is part of pBR322, growth in the presence of ampicillin forces the maintenance of pBR322 and also selects for the covalently attached tetracycline resistance gene. If the blue and white colonies obtained on the ampicillin plate containing X-gal and IPTG are transferred to a nutrient plate containing tetracycline at a concentration of 20 µg/ml, the blue colonies containing pUC19 will be unable to grow and the white colonies containing pBR322 will be able to form colonies. This is the fundamental principle that allows a plasmid or a viral DNA molecule to be used as a vector to carry and propagate exogenous, or extra, DNA fragments. DNA fragments that are covalently inserted into a vector by the use of restriction enzymes and DNA ligase will be maintained as the carrier vector molecule replicates in a host cell.

Materials

- Three nutrient agar plates containing 100 µg/ml ampicillin (NA+Ap plates).
- One nutrient agar plates containing 20 µg/ml tetracycline (NA+Tc plates).
- A water bath or temperature block at 43°C.
- Incubator for plate culture (this can be performed at room temperature, but growth of colonies will take two days).
- Glass spreader bar.
- Ethanol or isopropanol for sterilizing glass spreader bar.
- One 0.2 ml tube of competent JM83 (store these in a normal freezer [not frost-free if possible] until needed, then on ice).
- Ten µl of pUC19/pBR322 mixed plasmid DNA at 100 µg/ml.
- 2% X-gal in dimethylformamide (this solvent is necessary for X-gal).
- 100 mM IPTG in sterile water.
- Sterile toothpicks.

Procedure

1. Since X-gal and IPTG are not usually added to nutrient plates when the plates are liquid, it will be necessary to add these compounds to the surface of the solid plates prior to use. Place 0.1 ml of the X-gal in a 1.5 ml microcentrifuge tube (**not** a polystyrene tube - the X-gal solvent may melt the plastic). Add 20 µl of IPTG to

the tube and mix the two solutions to make an X-gal/IPTG mixture.

2. Place 60 µl of the X-gal/IPTG mixture on the surface of each of two of the plates containing ampicillin (NA+Ap). Save one NA+Ap plate and the NA+Tc plate for later use. Dip a glass spreader in alcohol and then ignite the alcohol in a burner flame to sterilize the spreader. Cool the spreader by touching the agar surface, then spread the X-gal/IPTG evenly across the surface of one of the NA+Ap plates. Again sterilize the spreader and spread the X-gal/IPTG mixture on the second NA+Ap plate. Invert the plates and label "+Xgal/IPTG". This step allows the coloring agents to soak into the surface of the agar. The agar surface may cloud as the dimethylformamide solvent dissipates and the X-gal precipitates. It is important to add the X-gal/IPTG mixture shortly before the plates are used because the X-gal can break down and may not color well if too old.

3. Place a tube of competent cells on ice and allow to thaw. Add the 10 µl of mixed plasmid DNA into the thawed cells and gently mix in. Return the cells to the ice.

4. Allow the tube to stand in the ice bucket for 15 to 20 minutes to allow the DNA to stick to the surface of the competent cells.

5. To make the DNA enter the cells, place the tube in a 43°C water bath (be careful with the temperature, too hot or too cool will not work as well). Allow the tube to incubate at 43°C for 60 seconds, then remove from the water bath and place at room temperature.

6. The transformation is complete. It is now necessary to plate the transformation mix on selective plates to detect the transformants and inhibit the growth of the non-transformed cells. Transfer 20 µl of the transformed cells to one nutrient plate containing ampicillin, IPTG, and X-gal. Use a sterile spreader to distribute the cells. Label the plate "20 µl". Spread the remainder of the

transformed cells on the second plate and distribute uniformly with a sterile spreader. Label the plate "180 µl".

7. Incubate the plates overnight at 37°C to allow growth of colonies. Check the plates for colonies after 18 hours growth and transfer the plates to a refrigerator for storage if the colonies are sufficiently large (the size of a typewritten"o") or are extremely numerous (>300 per plate). Although the colonies will stop increasing in size, blue color will continue to develop in the cold. If allowed to remain at 37°C too long, non-transformed, ampicillin-sensitive "feeder" colonies will begin to form as the ampicillin-resistant colonies degrade the ampicillin in the medium.

8. If the transformation has been successful, both blue and white colonies should be present on the surface of the two plates. The blue colonies are formed as a result of the transformation of bacteria with pUC19 and the white colonies are formed as a result of the transformation of bacteria with pBR322. This can be verified by using the ability of pBR322 to allow growth in the presence of tetracycline. To check transformants for this phenotypic property, label the back of the remaining NA+Ap plate and the NA+Tc plate with the numbers "1" to "40" and your initials as indicated below:

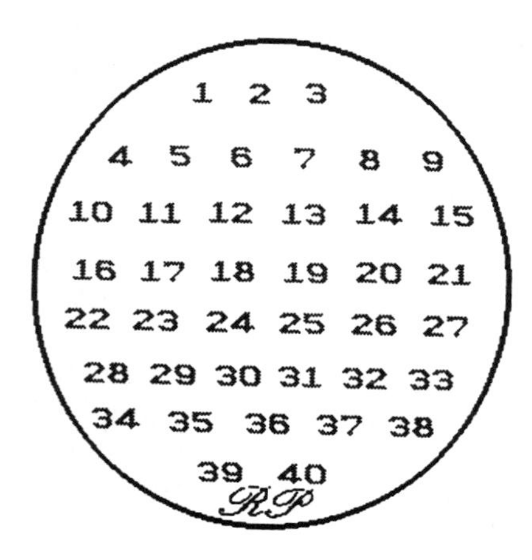

9. Touch a sterile toothpick to a blue colony on one of the NA+Ap X-gal/IPTG plates and make a short streak of cells over the number "1" on the NA+Ap plate and the NA+Tc plate. Repeat this for an additional 19 blue colonies and for 20 white colonies. Incubate both plates overnight at 37°C to allow the growth of the bacteria.

Analysis and significance of results

Since both the blue and the white colonies were originally obtained following growth on an NA+Ap X-gal/IPTG plate, all of these colonies should again grow on the NA+Ap plate but, since X-gal and IPTG are no longer present, all colonies should remain white. Only the white colonies obtained from the NA+Ap X-gal/IPTG plate should form colonies on the NA+Tc plate, confirming that the transformants that originally formed white colonies on NA+Ap X-gal/IPTG contained pBR322. Note that although the original selection for transformed bacteria involved resistance to ampicillin, resistance to tetracycline was also maintained by nearly all of the ampicillin-resistant white colonies. Since the genes conferring resistance to ampicillin and tetracycline are both present on the same DNA molecule (the plasmid pBR322), selection for one of the two genes (ampicillin resistance) also selects for the presence of the non-selected gene (tetracycline resistance).

It is possible to obtain ampicillin-resistant white colonies that cannot grow on tetracycline and ampicillin-resistant blue colonies that can grow on tetracycline. When the DNA from the plasmids present in these colonies is isolated and examined by digestion with restriction enzymes, the first type of colony can often be demonstrated to be caused by the occurrence of a deletion that removes part or all of the ß-galactosidase gene of pUC19, causing inability to metabolize X-gal to a blue compound, or a deletion that removes part of the tetracycline resistance gene, causing loss of tetracycline resistance. These deletions occur naturally in bacteria as an aspect of normal DNA replication, recombination, and repair processes.

The second type of colony is most commonly caused

by the presence of bacteria containing both pUC19 and pBR322 in the same colony. This can be verified by streaking the bacteria out on an NA+Ap X-gal/IPTG plate and incubating to obtain isolated colonies that are either blue and tetracycline-sensitive or white and tetracycline-resistant. While both plasmids can occassionally be inserted into the same bacterial cell during the transformation process, pUC19 and pBR322 replicate by very similar mechanisms and are said to be incompatible with one another. When incompatible plasmids are inserted into the same bacterium and the cell is grown in the absence of selective pressure designed to force maintenance of both plasmid types, the plasmids will segregate during division of the transformed bacterium and the resulting progeny will contain one or the other of the two plasmids.

Exercise 6. Comparison of plasmid and bacteriophage cloning vectors

Much of the scientific utility of recombinant DNA technology is based on the use of self-replicating DNA elements to serve as carriers for DNA fragments that are not capable of replication in bacteria. Bacterial cloning vectors are based on either bacterial plasmids or bacteriophage. Plasmids are naturally occurring, circular DNA molecules that can replicate as mini-chromosomes in bacteria. When isolated from nature, these DNA molecules also generally encode resistance to one or more antibiotics and may contain genes that allow transfer of the plasmid DNA from one cell to another during bacterial mating. Most cloning vectors derived from plasmids retain the plasmid origin of DNA replication and one or more antibiotic resistance genes, such as the *ß-lactamase* gene, which allows sensitive bacteria to form colonies on solid medium containing ampicillin. With plasmid cloning vectors, selection of bacteria containing the vector requires selection for a phenotype, or physical characteristic, associated with a gene present on the vector DNA molecule.

Cloning vectors have also been constructed from bacteriophage, viruses that replicate in bacteria. Two types of bacteriophage cloning vectors, derived from either the double-stranded, linear DNA bacteriophage lambda or the malespecific, single-stranded, circular DNA bacteriophage M13, are in common use in *E. coli*. Detection of cells containing either of these types of vector does not require selection for a phenotypic marker, but is based on the ability of the bacteriophage to form a plaque in a lawn of host bacteria, a property associated with the replication of the viral DNA vectors.

Lambda has a large genome (approx. 50 kilobase pairs of double-stranded DNA) that can, on insertion into a bacterial host, activate either of two sets of viral genes. One set of genes allows the bacteriophage to undergo a lytic cycle during which progeny phage are produced and the infected bacterial cell bursts open, releasing the daughter phage into the medium. When activated, the other set of genes causes the bacteriophage DNA to lysogenize, or insert itself into a specific location in the bacterial genomic DNA, where it remains silent while the bacterial cell undergoes normal cellular functions. Certain conditions, such as DNA damage, will induce the lysogen, activating the lytic genes. The bacteriophage DNA then excises from the host chromosome, produces progeny, and lyses the host cell. When bacteria are grown on the surface of solid medium from a low density to a uniform lawn of cells, the presence of lambda can be detected by the appearance of plaques, circular clear zones in the cloudy lawn of host bacteria. As the host bacteria grow to form the indicator lawn, successive cycles of infection/production of progeny/cell lysis cause the death and lysis of all of bacteria in a circular zone surrounding even a single lambda bacteriophage. Cells that are able to lysogenize the lambda DNA into their own genome become resistant to infection and may form small colonies within the zone of lysis. The ability of lambda DNA to cause the lysis of bacteria serves as the selective marker for the presence of cloning vectors derived from this bacteriophage. In marked contrast to plasmid cloning vectors, which require phenotypic selection to detect cells containing the vector, the replication of a lambda cloning vector provides a plaque, the immediate means of detection of cells containing the vector. Because the lambda plaque is basically a zone of lysed cells, however, the plaque indicates only where the infected cells were. Lambda vectors and recombinants derived from them must generally be maintained not as viable bacterial strains, but as stocks of virus particles.

An M13 bacteriophage particle contains a single-stranded, circular DNA genome of about 8 kilobase pairs. These particles will only infect host bacteria that are producing the F-pilus protein, which is part of the mating apparatus encoded by the large F plasmid. Bacteria that are male, or F^+, can be infected by the M13 phage particle. The entering single-stranded viral DNA is converted to and replicates as a double-stranded circular DNA. Single-stranded progeny phage identical to the original infecting DNA are produced, coated with viral coat proteins, and extruded through the bacteria cell wall into the medium without lysing the host cell. When propagated in a sensitive host grown on the surface of a solid medium, cloning vectors derived from M13 will form small, circular zones of infected cells that are extruding phage particles. In contrast to plaques formed by lambda vectors, these plaques are cloudy rather than clear and contain viable, infected cells that can be propagated as bacterial strains. While these plaques are biologically quite different than those induced by strains of lambda, the plaques appear similar to those induced by lambda. M13 vectors and recombinants derived from them can be maintained as either viable infected bacteria containing the double-stranded circular replicating form of virus DNA or as stocks of viral particles containing single-stranded circular DNA.

This exercise will use JM101, an F^+ strain of *E. coli* that will allow growth of plasmids and M13 to illustrate the similarities and differences in the use of these different types of cloning vector.

Materials

- Six tubes competent JM101 bacteria.
- 11 μl pUC19 DNA, 100 μg/ml (Amp^r plasmid vector).
- 11 μl mp19 DNA, 100 μg/ml (M13 cloning vector).
- Two LB plates containing 100 μg/ml ampicillin (LB-Ap).
- Two LB plates (LB).
- Twenty-five ml LB-soft agar (LB containing 0.8% agar).
- Sterile 5-10 ml glass or plastic tubes.

Procedure

1. Place the tubes of competent cells in an ice bucket to thaw. Label the two LB-Ap plates "pUC19 1 µl" and "pUC19 10 µl". Label two LB plates "mp19 1 µl" and "mp19 10 µl". Leave the plates at room temperature to warm up for about 10 minutes.

2. When the competent cells have thawed, label the tubes as below:

Tube 1	pUC19 1 µl
Tube 2	pUC19 10 µl
Tube 3	mp19 1 µl
Tube 4	mp19 10 µl

3. Add the appropriate amount of each of the DNA samples to the labeled tubes of competent cells and gently mix in. Allow to incubate on ice for 20 minutes.

4. Label four of the capped sterile glass or plastic tubes as below:

Tube 1	pUC19 1 µl
Tube 2	pUC19 10 µl
Tube 3	mp19 1 µl
Tube 4	mp19 10 µl

5. Use a microwave or boiling water bath to melt the 25 ml of LB-soft agar. Place the melted soft agar in a 55°C bath for at least 5 minutes to allow the temperature to decrease from boiling. Use a sterile pipet to dispense 3 ml of the liquid agar into the tube labeled "Tube 1 pUC19 1 µl". Immediately use a pipet device and sterile tip to transfer the competent cell/DNA mixture in the tube labeled "pUC19 1 µl" into the 3 ml of liquid agar. Pour the mixture onto the surface of the LB-Ap plate labeled "pUC19 1 µl" and gently rotate the plate to distribute evenly over the surface of the plate. Carefully place the overlaid plate aside to solidify and **do not move it for 5 minutes**. If this process is too slow, the agar will solidify

in the tube or in lumps on the surface of the plate. If the overlaid plate is inverted too soon after the pouring process, the soft agar will slide off of the agar plate.

6. Repeat this overlay for each of the remaining tubes of competent cells.

7. Incubate the plates at 37°C to allow growth of the bacteria.

Analysis of results

Following 12-18 hours of incubation at 37°C, the bacteria will have grown sufficiently to compare the difference in the two types of vectors. For the cells transformed with pUC19, the ampicillin in the LB plate rapidly diffuses into the soft agar overlay and retards the growth of the non-transformed cells. Cells that contain the plasmid are resistant to the antibiotic and form small colonies in and on the surface of the soft agar overlay.

For each of the bacteriophage vector transformations, the non-transformed bacteria will have formed a fairly uniform layer of cells called a lawn. The M13 mp19 vector will cause numerous holes and depressions in the lawn. These are best viewed by holding the plate up to a bright light.

Note that as the size of the vector increases, the number of transformants decreases (# pUC># mp19). This is principally caused by the difference in the sizes of these vectors (about 3 kb and 8 kb).

Exercise 7. The miniscreen: Rapid isolation of plasmid DNA

The construction of a library of recombinant plasmids involves the digestion of chromosomal and vector DNA with a restriction enzyme, followed by joining chromosomal DNA fragments to the linearized vector DNA with the enzyme T4 DNA ligase, followed by insertion of the DNA into a host cell. Because the presence of restriction enzymes can interfere with the ligation process, the digested DNA samples are generally extracted with phenol/chloroform to remove protein. Then, because ligation activity is not optimal in the buffer used for restriction of the DNA's, samples are precipitated with ethanol, washed with 70% ethanol to remove salts, dried, and resuspended in ligation buffer. The enzyme T4 DNA ligase, which uses ATP as a co-factor, is added to seal the nicks in the DNA molecules and generate recombinant molecules. The ligated DNA sample can then be used to transform competent cells to give blue and white colonies.

The blue and white colonies obtained after transformation of cells with a ligated DNA sample can be cultured and the plasmid DNA present in the cells extracted and examined. When purifying circular plasmid DNA from a bacterial cell, it is necessary to separate the plasmid away from the chromosomal DNA. Differences in the physical properties of these two types of DNA facilitate this separation. While purification of large (milligram) amounts of plasmid can be somewhat time-consuming and involve the use of large cultures (typically 1 liter) of bacteria, a variety of rapid isolation procedures allow the purification of a small amount of plasmid DNA (1- 5 μg) from a small culture (1-10 ml). The DNA obtained by these small-volume, rapid-isolation methods, generally referred to as "miniscreens" or "minipreps", is not as pure as that obtained by more elaborate methods, but is suitable for many characterization and cloning procedures.

These miniscreen procedures arose from the need to quickly sort through a number of bacterial transformants that contained recombinant DNA molecules of potential interest. A typical gene isolation experiment might generate 20 candidates for the desired gene construction. The time, expense and labor involved in purifying the DNA from a 1 liter culture of each candidate were found to be quite annoying, and abbreviated protocols were developed to allow the simultaneous screening of many candidates.

Rapid miniscreen procedures take advantage of the different purification properties of chromosomal DNA and supercoiled plasmid DNA to obtain a partial purification of the plasmid from a small culture. Steps of the process are generally performed in 1.5 ml plastic tubes, and centrifugation is in a microcentrifuge that can spin 12-24 tubes at 12,000 to 16,000x gravity. Spins used to remove precipitates are on the order of 5-15 minutes, and the entire miniscreen process will take about 90 minutes to purify the plasmid DNA from 12 samples. These protocols facilitate both the mass screening of transformants containing DNA molecules of potential interest and the rapid analysis and manipulation of DNA samples of particular interest.

Following transformation of JM83 with the pUC19:*E. coli* library DNA, incubation of the LB-Ap plates for 12-18 hr should result in the formation of bacterial colonies on the surface of the plates. Those transformants that contain only pUC19 are both Ap^r and able to metabolize X-gal to a blue compound and form blue colonies, while those cells that contain pUC19 plasmids with DNA inserts are Ap^r but unable to metabolize X-gal and therefore form white colonies. The ratio of white to blue colonies is an indicator of the frequency of recombinant transformants in the population.

To characterize the plasmids in the recombinant transformants, white colonies are picked with a sterile loop and transferred into 2 ml cultures of LB medium, allowed to grow for 6-18 hr at 37°C with shaking, then subjected to a miniscreen procedure. These cultures can either be inoculated the day before needed or grown several days in advance and stored in a refrigerator until needed. Long term storage of cells for DNA extraction is best accomplished by harvesting the cells from the cultures (step 1 below), adding the SET buffer, and freezing the cell pellets. To proceed with the miniscreen, thaw the pellets and vortex to resuspend the cells, then resume with step 3.

Materials

- LB medium: 1% Bacto Tryptone, 0.5% Bacto Yeast Extract, 0.5% NaCl, pH 7.0-7.2 (sterilized) for propagation of cultures to be miniscreened. Any complete nutrient medium can be substituted.
- One 2 ml LB *E. coli* JM83 culture, grown at 37°C with shaking for 8-15 hours.
- One 2 ml LB *E. coli* JM83 culture containing the plasmid pUC19, grown at 37°C with shaking for 8-15 hours.
- Four 2 ml LB *E. coli* JM83 cultures from white recombinant colonies obtained from the pUC19:*E. coli* library, grown at 37°C with shaking for 8-15 hours.
- SET buffer: 20% sucrose, 50 mM TRIS-HCl pH 7.6, 50 mM EDTA.
- Lytic mix: 1% SDS, 0.2 N NaOH.
- Sodium (Na) acetate: 3.0 M, pH 4.8. Make 3 M acetic acid and 3 M Na acetate, then mix to pH 4.8. If not made this way, miniscreens will not necessarily work well. Store at 4°C.
- RNase stock: Pancreatic ribonuclease A (RNase A), 1 mg/ml in 0.1 M sodium acetate, 0.3 mM EDTA.
- Isopropanol: Room temp.
- Ethanol: 70%, room temp. (70% isopropanol can be substituted).
- Water (dH_2O): Sterile, deionized or distilled.
- Ice.
- Conical microcentrifuge tubes and pipet tips.

Flow chart, Exercises 7 and 8

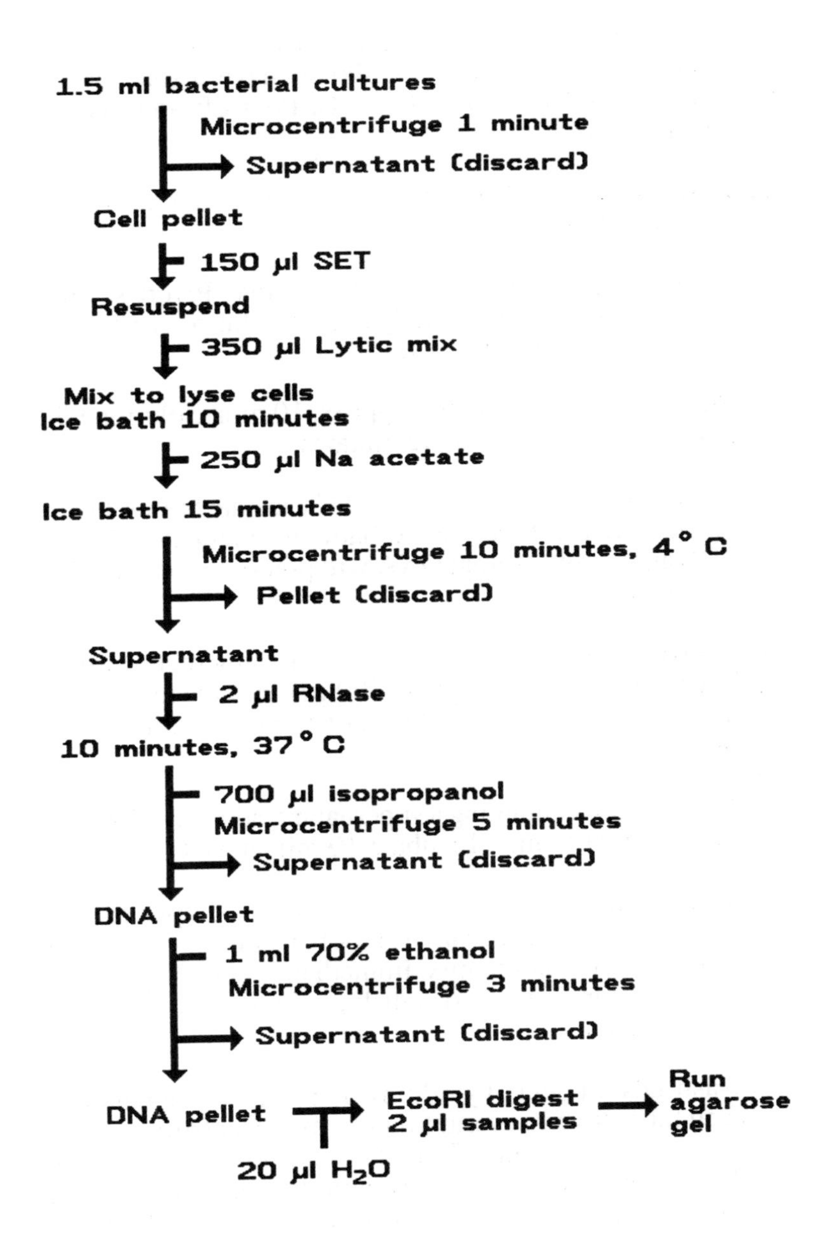

Protocol

1. Transfer 1.5 ml of each culture to a labeled microcentrifuge tube. You should have a total of six tubes. Spin 1 min to pellet cells. Pour off and discard supernatants.

2. Resuspend cell pellets in 150 µl of SET buffer. Vortex or agitate to resuspend cells.

3. To each tube add 350 µl of Lytic mix. Invert several times to mix. Cells will lyse and solution will clear slightly. Viscosity increases.

4. Place in ice bath and chill approx. 10 min. Solution will begin to cloud as SDS precipitates.

5. To each tube, add 250 µl of cold Na acetate buffer. Invert to mix. Return tubes to ice bath and incubate 15 min. SDS and chromosomal DNA will precipitate during this incubation.

6. Centrifuge tubes 10 min at 4°C in a microcentrifuge. Pour supernatants (approx. 700 µl) into clean, labeled microcentrifuge tubes. Discard tubes containing the pellets.

7. To each tube containing a supernatant, add 2 µl of RNase stock. Invert to mix. Incubate 10 min at 37°C.

8. Add an equal volume (approx. 700 µl, simply fill the remainder of the space in the tube) of isopropanol. Invert tubes several times to mix. Immediately centrifuge 5 min at room temp in a microcentrifuge. Pour off and discard supernatants.

9. Wash DNA pellets by adding 1 ml of 70% ethanol to each tube. Invert several times to mix. Centrifuge 3 min at room temp. Pour off ethanol and use a Kimwipe to dry the lip of each tube. Vacuum dry the DNA pellets. Resuspend each pellet in 20 µl dH_2O. After DNA has

been allowed to resuspend for 10 minutes on ice, tap the tubes gently to help resuspend the DNA, then centrifuge 20 seconds to collect solution in the bottom of the tubes. The DNA is now ready to be digested with restriction enzymes or can be stored frozen until further use.

Exercise 8. Characterization of pUC19 recombinant DNAs

The miniscreen DNAs will now be digested with restriction endonuclease *Eco*RI, the same enzyme that was used to construct the library. Any DNA fragments that were inserted in the unique *Eco*RI site of pUC19 will be released upon digestion, and agarose gel electrophoresis will allow characterization of the size and number of additional fragments.

Materials

- Miniscreen DNAs.
- Ice.
- Conical microcentrifuge tubes and pipet tips.
- Endonuclease digestion materials, see Exercise 2.
- Gel electrophoresis materials, see Exercise 1.

Protocol

1. If you plan to digest more than a few samples, it is convenient to make a mix containing 10x *Eco*RI buffer, H_2O, and *Eco*RI enzyme, aliquot this mix into reaction tubes, then to each tube add a separate DNA sample. For example, for each six miniscreens to digest, plan on a 20 µl reaction for each digest, with 2 µl of miniscreen DNA diluted into 18 µl of reaction mix. Prepare the following mix:

10x *Eco*RI buffer	12 µl
dH_2O	96 µl
*Eco*RI enzyme, 20 units	2-5 µl
Total volume	110-113 µl

Label six tubes for the digestion reactions, and dispense 18 ul of this mix into each tube. Then add 2 µl of a miniscreen DNA to each of the tubes, for a total of six different reactions. This is more convenient than setting up six individual reactions.

2. Incubate at 37°C for 30-60 min, then remove 8 µl from each digestion and place on a Parafilm strip. Add 3 µl of SM dye to each sample.

3. Subject samples to electrophoresis on a 1% agarose gel in TBE buffer. Run a sample of *Eco*RI- or *Hin*dIII-digested lambda DNA as a molecular weight standard.

4. Following electrophoresis, staining and visualization of the DNA fragments will reveal any additional fragments present in the pUC19 recombinants. While a photographic record of the gel allows convenient measurement of fragment position for the purpose of calculating mobilities, and hence determining molecular weights, photos are expensive. A ruler can be laid next to the gel on the UV box, and a drawing with measurements added used to construct a graph of fragment mobilities.

Analysis and significance of results

A considerable amount of sample variability is common when learning to prepare miniscreen DNA. A typical gel with good quality DNA is illustrated below:

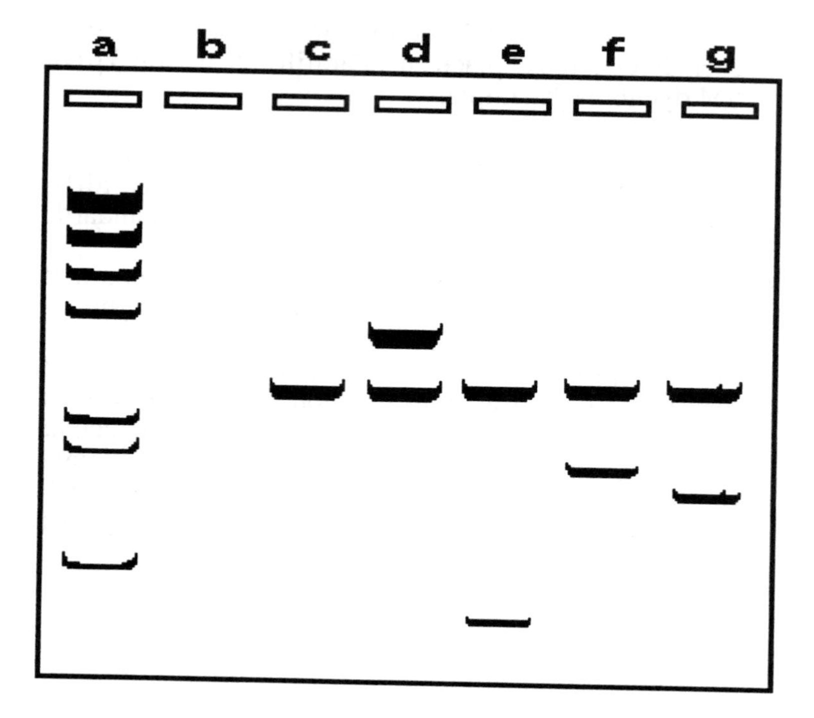

Lane a. *Hin*dIII lambda DNA
Lane b. JM83 (no plasmid)
Lane c. JM83 (pUC19)
Lane d. JM83 (recombinant #1)
Lane e. JM83 (recombinant #2)
Lane f. JM83 (recombinant #3)
Lane g. JM83 (recombinant #4)

Note that no specific DNA fragments other than a faint background of chromosomal DNA fragments are observed in the miniscreen DNA prepared from the JM83 culture containing no plasmid, while the five samples prepared from Ap^r transformants all contain specific DNA bands. This further demonstrates that the presence of the plasmid DNA in the transformants is associated with the conversion to ampicillin resistance. In addition, the presence of inserted DNA fragments correlates with the inability of recombinant pUC19 plasmids to generate blue transformant colonies. Different recombinant colonies may contain inserts of different sizes. Small inserts may not be detected by agarose gel electrophoresis.

Many of the gel artefacts commonly associated with miniscreen DNA are illustrated below:

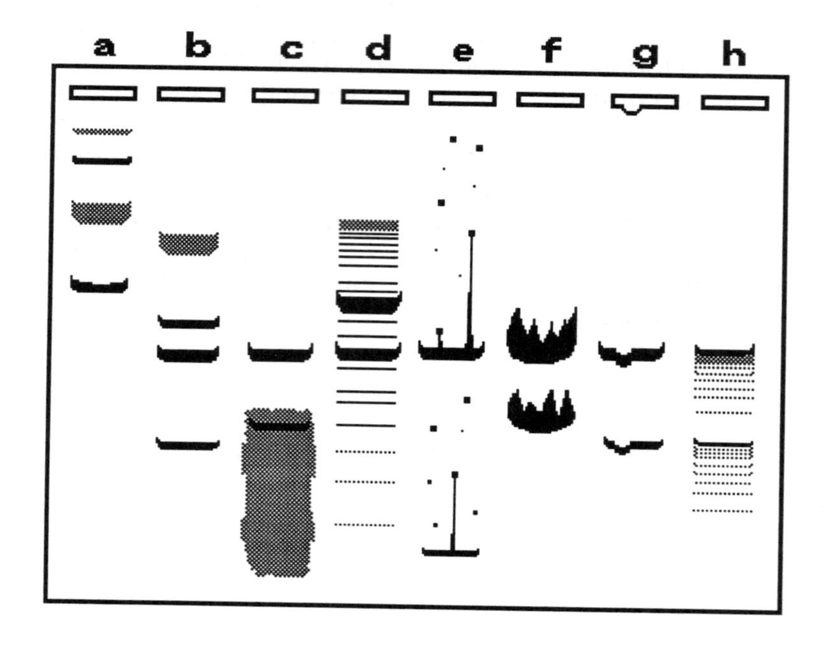

Lane a. Undigested DNA with both supercoiled (dark bands) and open circular (stippled bands) forms.
Lane b. Partially digested plasmid DNA sample with insert (bottom band), plasmid vector (second from bottom), linear recombinant cut at only one of two cleavage sites (third from bottom), and open circular DNA (stippled band).
Lane c. Digested plasmid DNA sample containing undigested RNA (stippled region).
Lane d. Digested plasmid DNA sample contaminated with digested bacterial chromosomal DNA not eliminated during the DNA preparation.
Lane e. Unmelted agarose specks causing DNA streaking and spots in the gel.
Lane f. Too much DNA loaded in the lane causing distortion of bands.
Lane g. Bubble in sample well deforming DNA bands.
Lane h. Digested DNA sample degraded by contaminating exonuclease or endonuclease.

Exercise 9. DNA fingerprinting

The ability to digest DNA fragments, separate these fragments according to size, and detect fragments containing specific genes has led to a genetic identification approach called DNA fingerprinting. This method takes advantage of the fact that a single nucleotide sequence difference can cause the appearance or disappearance of a restriction enzyme cleavage site in DNA. The individual-specific differences in restriction enzyme cleavage site position relative to a specific gene can cause variation in the size of the DNA restriction fragment that carries that gene in different individuals. The variation in the size of a restriction fragment that carries a specific gene, referred to as "restriction fragment length polymorphism" (RFLP), can be used to genetically match DNA samples with the donors from which they were obtained. Individuals who appear identical for a specific physical trait can be easily distinguished if sequence differences that cause changes in DNA restriction fragment sizes are associated with that trait.

When applied to identification of human DNA samples, DNA fingerprint analysis is complicated by the presence of hundreds or thousands of DNA fragments following digestion of DNA samples with a specific restriction enzyme. Rather than attempt to analyze all of the DNA fragments, the fragments are generally separated by gel electrophoresis, transferred to a membrane, and allowed to react with a DNA or RNA hybridization probe that will specifically anneal to a small number of the many different DNA fragments. The probe fragment generally contains a nucleoside that is either radioactive (such as ^{32}PdATP) and can be detected by autoradiography or has been modified (such as biotinylated dATP) to be recognized by specific antibodies and detected by a color reaction. The image that is obtained by visualization of the probe is a subset of the total set of DNA fragments and can be referred to as a DNA fingerprint. When hybridization probes are chosen carefully, the fingerprints can be specific for DNA that has been found to have a high degree of individual-specific variation. When sufficient information has been obtained about the frequency of occurence of a specific DNA fingerprint pattern, the DNA fingerprints of two DNA samples can be used to calculate the probability of whether the two DNA samples were obtained from the same individual.

Since humans are diploid and contain two copies of each chromosome, one from the mother and the other from the father, a human DNA fingerprint can be thought of as the combination of two fingerprint patterns. Since these patterns can be used to link an "unknown" DNA sample to DNA samples obtained from specific individuals, DNA fingerprint analysis is becoming increasingly common as courtroom evidence in paternity suits and in criminal prosecution involving crimes like rape and murder. This exercise will use bacteriophage lambda DNA digested with various restriction enzymes to simulate the results of a DNA fingerprint analysis and help determine paternity of a child.

Materials

Bacteriophage lambda DNA at a final concentration of 50 μg/ml in gel loading buffer, digested with the following restriction enzymes: *Hind*III (H), *Eco*RI (E), *Bam*HI (B), *Kpn*I (K), and *Sal*I (S). Each of these digests will be used to simulate a haploid DNA fingerprint (one set of chromosomes). To make the test DNA samples obtained from the individuals involved in the paternity suit, the digested lambda DNA samples have been mixed in the following manner to simulate the diploid chromosomal fingerprint of each individual:

Mother = 100 μl H + 100 μl H (HH)
Baby = 100 μl H + 100 μl E (HE)
Rock Star = 100 μl H + 100 μl K (HK)
Ex-drummer = 100 μl S + 100 μl B (SB)
Mailman = 100 μl S + 100 μl E (SE)

Loading 7 μl of each mix will deliver 350 ng of DNA, giving distinctive diploid DNA fingerprint patterns for each individual.

- Gel box, gel tray and comb.
- Power supply.
- UV light source.
- Dry agarose.
- 10 mg/ml ethidium bromide.
- 10X TBE gel buffer.

Background

While living with a rock star, a young woman becomes pregnant, whereupon the rock star promptly severs his relationship with the woman, leaving her and the resultant baby completely penniless. The mother of the child sues for child support, claiming the rock star as the father of her baby. The following are the summary statements from each of the individuals who become involved in the resulting paternity suit:

Mother: Rock Star is the father of my baby. He should be required to financially support us.

Baby: Wah! Burp.

Rock Star: The child is not my son. The males in my family were found through genetic analysis to carry a recessive gene that causes incurable craving for disco music. On the advice of my doctor, I had a vasectomy when I was twenty-three to prevent transmission of this genetic disorder to my children. I kicked her out because I caught her in bed with my Drummer, not because she was pregnant. She should talk to him about support.

Ex-drummer: Who says I'm the father? I wasn't the only other guy she was seeing. I even caught the mailman in the bedroom with her. I couldn't support her anyways. I can't get work since Rock Star kicked me out of his band and I got busted with 2 pounds of heroin.

Mailman: Sure I was in her bedroom. I had a package for Rock Star and needed a signature. When I rang the doorbell, she called out she was sick in bed, but the door was unlocked and she would sign for the package if I didn't mind risking the flu. I was leaving her bedroom when Drummer walked in the front door. I think the door was unlocked and she was in bed waiting for Drummer.

At the request of the court, DNA is extracted from blood samples obtained from each of these individuals and provided to you, who will perform the fingerprint analysis and interpret the results.

Procedure

1. Weigh out 0.5 gram of the agarose and place in a 100 ml flask or beaker. Add 45 ml of distilled water and 5 ml of 10X TBE buffer.

2. Melt the agarose completely in a microwave oven or boiling water bath. The solution must boil to melt the agarose. Agarose has a tendency to superheat and flash-boil, so be careful on the first boiling. The boiling will be more controlled once some of the agarose has melted. Be certain all the small particles are melted to prevent streaks in the gel. The agarose should be allowed to cool to about 55°C. It will solidify at about 45°C.

3. Add 2 µl of ethidium bromide to the agarose and mix. This is the stain to detect the DNA. Ethidium bromide is a mutagen (comparable to a pack of cigarettes) and

solutions containing ethidium should not be handled with bare hands.

4. Pour a gel as demonstrated and allow to solidify.

5. Assemble the gel in the gel box. Submerge the gel in 1X TBE running buffer made by diluting 10 ml of 10X TBE Buffer with 90 ml of distilled water.

6. Load the samples in the following order and amount:

Lane	DNA Sample	Amount
1	Mother	7 µl
2	Baby	7 µl
3	Rock Star	7 µl
4	Drummer	7 µl
5	Mailman	7 µl
6	Baby	7 µl

7. Plug in the leads and turn on the power. Voltage of around 80 to 120 volts is typical, but will vary depending on the gel box. A very low voltage will slow the run down, but will enhance resolution of the DNA bands. A very high voltage will heat the buffer and can actually melt the gel. The depth of the buffer determines how much current flows in the system. Use only enough buffer to just submerge the gel to prevent excessive heating of the gel (deep buffer = high current = high heat production).

8. As the run progresses, two blue bands will resolve in the gel. These are merely tracking dyes to judge how far the DNA has migrated. When the faster dye is reaches the bottom of the gel, turn off the power. Remove the gel from the box and expose to the UV light. Be careful to shield eyes and skin from the UV light, as it is both a mutagen and a skin and eye burning agent.
9. Either photograph or sketch the resulting band patterns. Analyze the DNA band patterns. Remember that the DNA fingerprint of each person will contain two sets of bands, one from each parent. These could be either two identical sets of bands, as in the case of the Mother (HH),

or two different sets of bands, as in the case of the Baby (HE). The Mother must have obtained one H set of bands from her mother and one H set of bands from her father. The Baby must also have obtained one set of bands from its mother and one set from its father.

Questions you will need to consider:

1. Is Rock Star the father of the Baby? Why or why not?

2. Is Drummer the father of the Baby? Why or why not?

3. Is Mailman the father of the Baby? Why or why not?

4. The probability that a DNA fingerprint will contain a particular DNA band pattern is dependent on the frequency of that particular band pattern in the general population. You are now provided with the information that the entire population of the USA has been screened with the DNA probes you have just used and the frequencies of occurrence of the H, E, K, B, and S band patterns have been determined to be as follows:

Band pattern	Frequency in population
H	1 in 100 people
E	1 in 10 people
K	1 in 100,000 people
B	1 in 100,000 people
S	1 in 50 people

   Does this information change your answers to questions 1-3 above, and why or why not?

5. As the court-appointed expert in DNA fingerprints, what is your recommendation to the court regarding the paternity suit and the identity of the father of the Baby?

Appendix I

Plans for simple submerged gel electrophoresis box

Pieces required to assemble the box and lid can be cut from 1/4 inch Plexiglas or comparable acrylic sheet. Cut edges should be rounded with a file or fire-polished to eliminate sharp corners. Electrode wires should be either platinum or gold (gold is less expensive) to minimize reactivity with electrophoresis buffers. Most other metals will corrode.

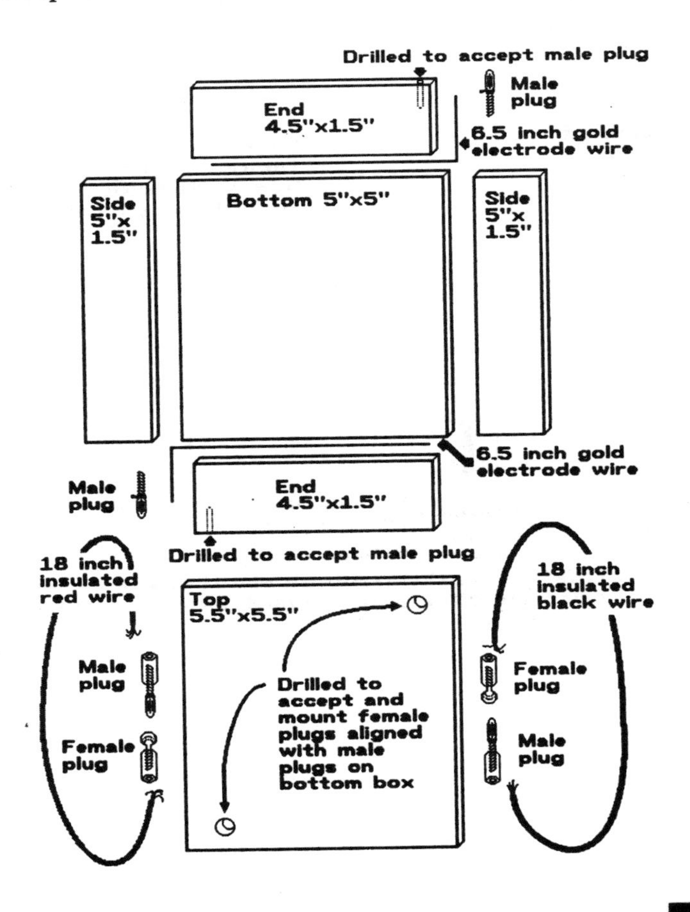

Plastic parts can be taped together and sealed with a Plexiglas glue or acrylic plastic cement. Be certain seams are well sealed and leak free. Once the glue has dried, attach the electrode wires at the ends of the box by holding the wire in place with tweezers and touching the wire with a hot soldering iron. The wire will heat and melt into the plastic for a permanent mount. Attach the wire at four or five points along the bottom of the box, then screw the male terminals into the holes drilled in the end piece and attach the electrode wires to the terminals. Coat the indicated portion of each electrode wire with silicone rubber to minimize excess current. If left uncoated, this excess electrode wire can cause uneven current in the gel box, resulting in uneven mobility of DNA samples across the width of the gel box.

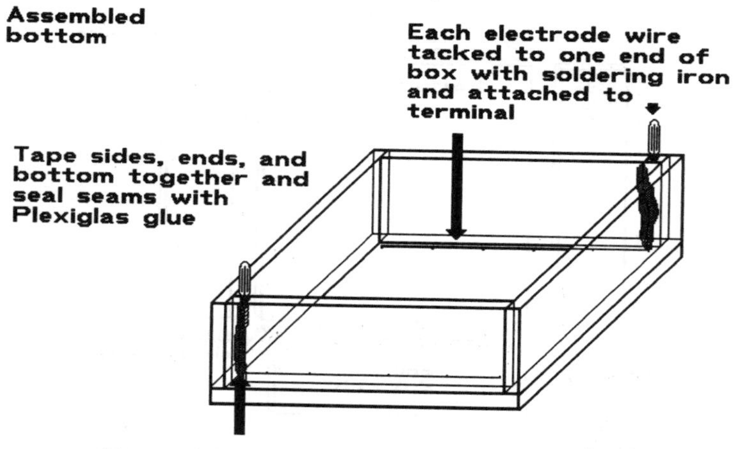

Lid assembly involves attaching the black and red wires to the appropriate color terminal plugs and mounting the female plugs in the holes in the lid.

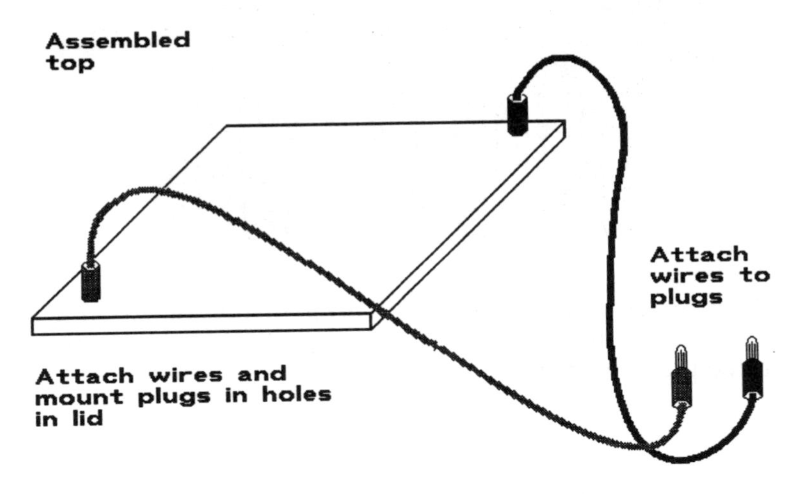

Because the power supply terminals are attached to the lid, the lid must be placed on the box to deliver current to the electrophoresis chamber. This safety feature minimizes chances of electrical shock in the event that the lid is removed from the box without first turning off the power supply.

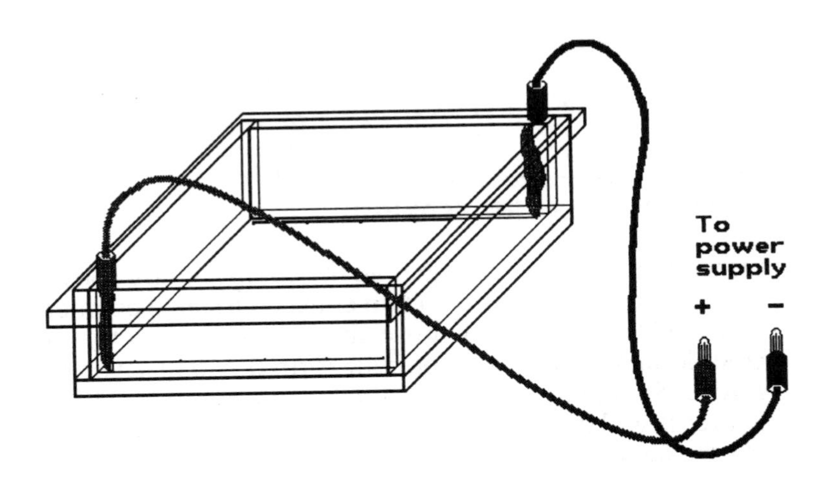

Slot-forming combs can be made from 1/16 inch plastic cut with a bandsaw or hacksaw.

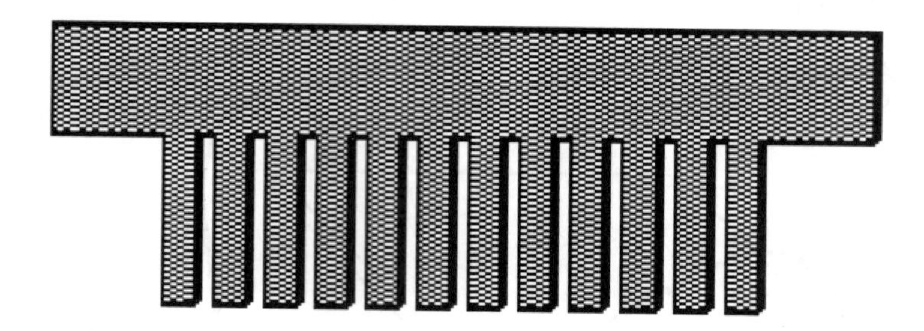

Agarose gels can be poured either on disposable 50 mm x 75 mm glass microscope slides or in a tray assembled from three pieces of plastic.

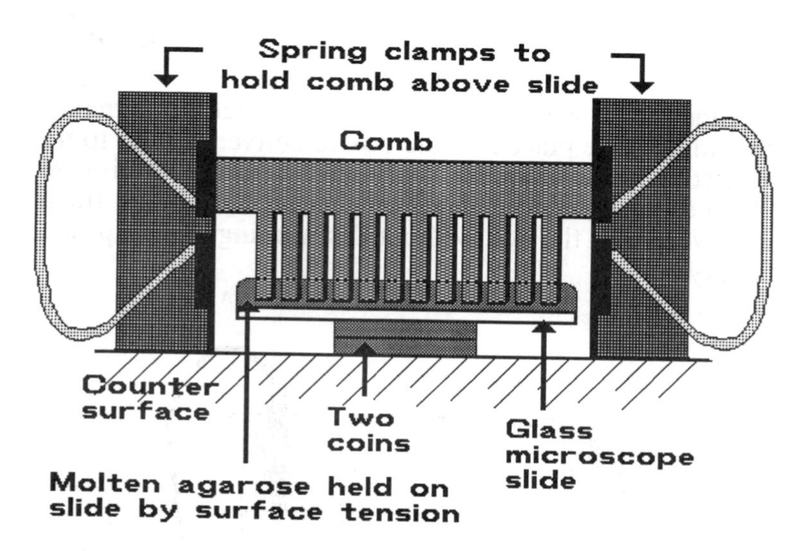

Spring clamps can be used to hold the comb slightly above a microscope slide placed on two coins on a stable counter surface. Cooled molten agarose can be poured directly on the slide and surface tension will form a 5 mm layer of agarose. Should too much agarose be layered on the slide, the excess will drip off without siphoning all of the agarose onto the counter.

Small trays can also be assembled from three pieces of plastic glued together. The size of the tray can be chosen to fit the specific gel electrophoresis boxes in use.

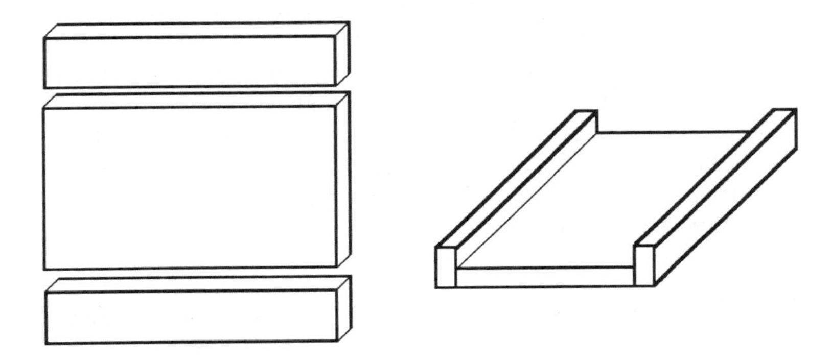

Gels are poured by placing tape over each end of the tray and using spring clamps to hold a comb slightly above the bottom of the tray, as illustrated below. The cooled molten agarose can be poured into the tray and allowed to gel. After removal of the comb from the gel, the gel can either be left in or removed from the tray and placed in the electrophoresis box.

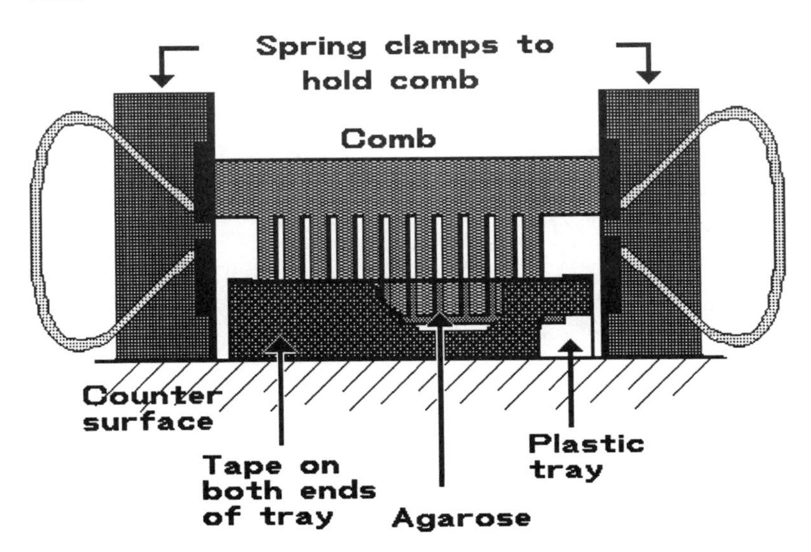

The prepared gel should be transferred to the gel box and covered with electrophoresis buffer so that the gel is covered by about 2-3 mm of buffer. Agarose gels are traditionally run with constant voltage of 80 to 150 volts. Too much buffer will increase the current flow during electrophoresis and increase heating of the gel. Excess heat can distort the migration of the DNA samples in the gel and can actually melt the gel during electrophoresis. If too much buffer is added to the electrophoresis unit, the voltage will have to be decreased to reduce the current and minimize excess heating. This reduced voltage will slow sample migration and increase the required run time.

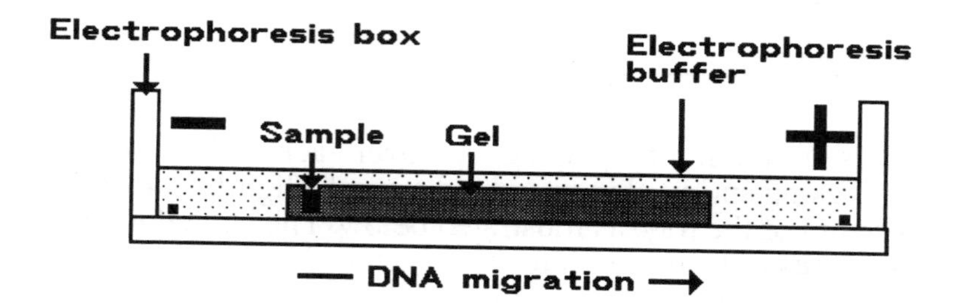

Appendix II

Manual pipet device

Most manual pipet devices that can accomodate samples less than 100 µl are relatively expensive ($150-250). The disposable plastic tips used with these devices are quite inexpensive and can be used with a disposable syringe as a gel loading device. The barrel of the syringe is too small to accept the plastic tip directly, but the tip can be trimmed with a razor blade as indicated to achieve a tight fit. The gradations on the side of the syringe are usually in .01 ml (10 µl) increments and can be used to estimate sample volumes. These syringe micropipets are convenient for loading gels, but are not accurate enough for setting up enzyme reaction mixes.

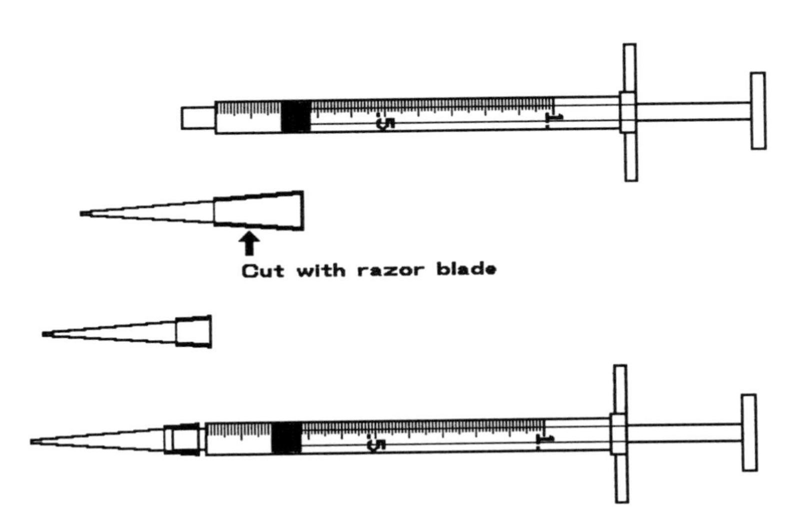

Appendix III

Bacterial DNA purification

Many commercial DNA preparations (calf thymus, salmon sperm, herring sperm) are not sufficiently free of contaminants to allow digestion with restriction enzymes. This protocol can be used to prepare a high-quality bacterial DNA that can be efficiently digested with most restriction enzymes. The bacterial cell wall will be weakened by treatment with the cell wall degrading enzyme lysozyme and with the chelating compound EDTA. Cells will subsequently lysed by the addition of the ionic detergent sodium dodecyl sulfate (SDS). Contaminating DNase that is released during cell lysis is inhibited by the EDTA, which binds the Mg^{++} ions required for nuclease activity, and by the SDS, a strong disrupting agent of protein structure and therefore an inhibitor of DNAse activity. The enzyme pancreatic RNAse is resistant to the inhibitory effects of EDTA and SDS and is added to degrade RNA, which would otherwise co-purify with the DNA and might interfere with subsequent quantitation and digestion of the DNA. Heat-treatment of the RNAse prior to use destroys contaminating DNAses. The protease pronase is added to help degrade proteins, then phenol/chloroform extraction used to remove proteins from the DNA. The resulting DNA will be concentrated by precipitation in ethanol, washed free of contaminating compounds, resuspended, and the resulting concentration determined.

Materials

- An 80 ml culture of *E. coli* bacteria, grown to late logarithmic phase in a 250 ml flask in a 37°C shaking incubator. Any rich medium will suffice for growth.

- SET buffer: 20% sucrose, 50 mM TRIS-HCl pH 7.6, 50 mM EDTA.

- RNase stock: Pancreatic ribonuclease A, 10 mg/ml, in 0.1 M Na acetate, 0.3 mM EDTA, pH 4.8, preheated to 80°C for 10 minutes once prior to use to destroy contaminating DNase, then stored at -20°C.

- TEN buffer: 10 mM TRIS-HCl pH 7.6, 1 mM EDTA, 10 mM NaCl.

- Phenol/chloroform: 1:1 mixture. Phenol is a crystalline solid at room temperatures unless it has been equilibrated with water. Solid phenol can be liquified by heating to about 50°C. Use a graduated cylinder to measure 100 ml of liquified phenol, then add 100 ml of chloroform. Carefully pour the phenol/chloroform into a brown bottle, cap tightly and gently swirl to mix the solution well. The addition of chloroform will both increase the density of the solution to improve phase separation during extraction of proteins and will keep the phenol liquid at 4°C.

- Chloroform.

- Lysozyme: 5mg/ml in TEN buffer.

- Pronase: 2 mg/ml in TEN buffer, preheated to 37°C for 15 minutes to degrade contaminating DNase, then stored at -20°C.

- Sodium Dodecyl Sulfate (Sodium Lauryl Sulphate, SDS): 25% in dH_2O.

- NaCl: 5 M.

- Ethanol: 95%, -20°C.

- Water (dH_2O): Sterile, deionized or distilled.

- Centrifuge and phenol-resistant tubes.

Protocol

1. Harvest the culture by centrifugation (Sorvall GSA rotor or equivalent, 250 ml phenol-resistant bottles, 5 minutes, 5000 rpm, 4°C). Pour off supernatants and drain cell pellet well. Pellet can be stored frozen for several days if needed.

2. To each frozen cell pellet, add 25 ml SET. Place tubes in warm water to thaw pellets. Vortex or use pipet to resuspend.

3. Add 1 ml of lysozyme and 0.1 ml of RNase stock. Incubate on ice 15 min.

4. Lyse cells by the addition of 0.5 ml SDS and 25 ml of dH_2O. Rotate tube to disperse SDS. Solution should become viscous. **Do not agitate harshly**, or DNA will be sheared and molecular weight reduced!!!

5. Add 2 ml Pronase. Incubate 37°C, 30-90 minutes.

6. Add 50 ml of Phenol/chloroform. Tighten cap and gently swirl tube for 5 minutes to mix the phases. **Caution: phenol burns skin rapidly!!** If spilled on skin, wash liberally with tap water. Wipe up spills on counters immediately or phenol will end up on arms and elbows.

7. Separate the two phases by centrifugation: 9krpm, 5 min, 4°C. Phenol/chloroform (protein containing) layer will be on bottom, aqueous (DNA containing) on top. A denatured protein pellicle is often observed at the interface of the two phases. If the cell pellet was resuspended in too small a volume of SET buffer, the protein pellicle will be very large and it will be difficult to withdraw the very viscous DNA solution.

8. Invert a 10 ml pipet and place a manual pipet device (such as a rubber bulb) on the tip of the pipet. Use the inverted pipet to remove the viscous upper aqueous phase and place in a clean 30 ml centrifuge tube

containing 50 ml of chloroform. Avoid the fluffy denatured protein when possible. Cap and again swirl the tube for 5 minutes to mix the phases. This chloroform extraction helps remove residual phenol from the DNA solution.

9. Centrifuge as in step 7. Again use an inverted pipet to withdraw the viscous aqueous phase and transfer to a clean tube. Repeat the chloroform extraction.

10. Again use an inverted pipet to withdraw the viscous aqueous phase and transfer to a clean 250 ml flask or beaker. To the final aqueous DNA solution (approx. 40 ml), add 1/25 vol of 5 M NaCl (final NaCl concentration = 200 mM). Swirl the solution to disperse the NaCl.

11. Add 2 vol of cold 95% ethanol (approx. 100 ml). Gently mix the solution. Allow to stand on ice 5 min. DNA will precipitate as a cottony white mass.

12. With a clean glass rod, such as the flame-sealed end of a disposable glass pipet, spool out the precipitated DNA. The DNA will attach to the glass rod and gentle winding or mixing with the rod will form a mass of DNA on the end of the rod. Appearance will be much like a cotton swab. Gently transfer the glass rod to a fresh tube of 95% ethanol. As the DNA dehydrates in the ethanol, it will contract and become more white. This rinse removes low molecular weight contaminants. Repeat the ethanol rinse 2 more times.

13. Air-dry the DNA on the rod for 5 minutes, then place the rod in a tube containing 2 ml TEN buffer. The DNA will begin to rehydrate and will detach from the rod. This is a slow process! Store at 4°C. High molecular weight DNA generally requires several hours, sometimes several days, to completely go into solution. The solubility limit of high molecular weight DNA is about 2 mg/ml. Stocks this concentrated can be difficult to work with due to extreme viscosity. It may be necessary to add more TEN to dilute the DNA as it goes into solution.

14. After dissolution is complete (for concentrated DNA solutions, this may required incubating overnight at 4°C), determine the concentration of the DNA. Take 0.05 ml of DNA solution and dilute into 1.95 ml TEN buffer. Measure the DNA carefully. Viscous stocks of DNA often "crawl out" of pipets during measurements. Using TEN buffer as a blank, determine the absorbance of the diluted DNA at wavelengths of 260 nm and 280 nm. Using the relationship 1 absorbance unit at 260 nm = 50 µg DNA/ml solution, calculate the concentration of the DNA samples. Multiply by 40 (the dilution factor) to determine the concentration of DNA in the stock tubes. Determine the ratio 260 nm/280 nm. With protein-free DNA samples, this ratio should be 1.9 to 2.0, depending on the source of the DNA. Discard the diluted DNA sample, as the UV light induces the formation of thymine dimers in the DNA, which can interfere with subsequent experiments.

15. You may wish to adjust the DNA stock to a specific concentration, such as 100 µg/ml. Chromosomal DNA stocks more concentrated than 200 µg/ml can be quite difficult to handle. If desired, overnight dialysis against a buffer like 10 mM Tris-Cl pH 7.6, 0.1 mM EDTA can be used to remove residual contaminants.

Appendix IV

Preparation of a pUC19:*E. coli* chromosomal library

Several of the exercises use a recombinant library to illustrate key points of recombinant DNA methodology. This library need not be of high quality to make the crucial points, and can be a sample of an actual library, a collection of miniscreen DNA samples that together comprise a small mixture of recombinant and non-recombinant molecules, or a ligation mix consisting of pUC19 ligated with *E. coli* chromosomal DNA. For routine student laboratory use, the most satisfactory results are generally obtained by providing aliquots of a mini-library consisting of a collection of pUC19 and several pUC19 recombinants with DNA inserts or diferent sizes.

A satisfactory ligation mix can be prepared by the following protocol. Transforming a bacterial host with the ligation mix and sorting through the blue and white transformants by miniscreen analysis of plasmid DNA can provide a collection of recombinants containing one or more DNA inserts of different sizes. Pooling leftover undigested miniscreen DNA samples will generate a mini-library of sufficient quality for student laboratory use.

Materials

- Pure *E. coli* chromosomal DNA; stocks are generally 100-500 μgrams per ml.
- pUC19 plasmid DNA, commercially available from several suppliers.
- *Eco*RI endonucelase and 10X digestion buffer.
- Water (dH_2O): Sterile, deionized or distilled.

- Agarose gel electrophoresis materials.

- Phenol/chloroform: 1:1 mixture. Phenol is a crystalline solid at room temperatures unless it has been equilibrated with water. Solid phenol can be liquified by heating to about 50°C. Use a graduated cylinder to measure 100 ml of liquified phenol, then add 100 ml of chloroform. Carefully pour the phenol/chloroform into a brown bottle, cap tightly and gently swirl to mix the solution well. The addition of chloroform will both increase the density of the solution to improve phase separation during extraction of proteins and will keep the phenol liquid at 4°C.

- Ethanol: 70%, room temperature.

- Ethanol: 95%, -20°C.

- T4 DNA ligase and 10X ligation buffer.

- Water bath and ice to maintain temperature at 12-16° C.

- Conical microcentrifuge tubes and pipet tips.

Protocol

1. Use the concentration of the chromosomal DNA stock to calculate the volume necessary to contain 20 μgrams of bacterial DNA. For a stock of 200 μg/ml, this would be 100 μl. Make an *Eco*RI digestion reaction in a labeled 1.5 ml conical tube by making the following additions:

chromosomal DNA	as calculated
10x *Eco*RI buffer	20 μl
dH_2O	to 200 μl final volume
Total volume	200 μl

2. Mix the contents of the tube gently but thoroughly. The chromosomal DNA should be quite viscous and difficult to handle. **Do not agitate harshly**, or DNA will be

sheared and molecular weight reduced!!! Add 50 units of *Eco*RI enzyme (generally 3-10 µl) and mix gently.

3. Mix, then incubate at 37°C for 30 minutes.

4. Set up a digestion reaction in a labeled 1.5 ml conical tube by making the following additions:

10x *Eco*RI buffer	10 µl
dH_2O	80 µl
1 ug pUC DNA	10 µl
Total volume	100 µl

5. Mix the contents of the tubes gently but thoroughly, then add 5-10 units of *Eco*RI enzyme, generally about 1 µl. Mix, then incubate at 37°C for 30 minutes.

6. Remove a 10 µl aliquot from each digestion reaction and monitor cleavage by subjecting samples to electrophoresis on a 1% agarose gel. Cleavage of the pUC19 DNA should be as complete as possible. If not complete, add additional enzyme and continue incubation for an additional 30 minutes. Chromosomal stocks that contain contaminants may not digest well, and dilution of reactions with an equal volume of 1X *Eco*RI reaction buffer prior to the addition of more enzyme will often improve extent of digestion.

7. When digestion has been determined to be sufficient, combine the two samples in a single microcentrifuge tube. Add an equal volume of phenol/chloroform and gently mix to inactivate and remove proteins. Spin in a microcentrifuge 5 min at maximum speed to form two phases.

8. Carefully remove the upper, aqueous phase that contains the DNA and transfer to a clean microcentrifuge tube. Be careful not to transfer any of the fluffy white interface that contains the inactivated proteins from the reaction.

9. Measure the recovered volume and add 2 volumes of cold 95% ethanol. For example, add 260 µl of ethanol to 130 µl of DNA solution. Invert the tube several times to mix the ethanol and precipitate the DNA.

10. Spin in a microcentrifuge 15 min at maximum speed to pellet the precipitated DNA. When the microcentrifuge stops, immediately remove the tube and pour off the liquid. A small DNA pellet should remain in the tube.

11. Fill the tube with 70% ethanol, cap and invert several times to wash the DNA pellet and remove residual phenol.

12. Spin in a microcentrifuge 15 min at maximum speed to pellet the precipitated DNA. When the microcentrifuge stops, immediately remove the tube and pour off the liquid. While holding the tube upside down, use a tissue to absorb residual ethanol. Place the tube on its side and allow to dry completely (most easily accomplished in a vacuum.

13. When the DNA pellet is dry, prepare a ligation reaction in the tube containing the DNA pellet by making the following additions:

10x ligase buffer	20 µl
dH_2O	180 µl
Total volume	200 µl

14. Mix the contents of the tube gently but thoroughly. The DNA pellet may resuspend somewhat slowly and can be placed at 42° C for 30 minutes.

15. When the pellet appears dissolved, place the ligation reaction tube in a water bath maintained at 12-16° C by adding small amounts of ice. This temperature will improve the rate of annealing of the cohesive restriction termini. Add 2 µl of T4 DNA ligase and allow to incubate in the water bath for several hours.

16. Ligation efficiency can be monitored by agarose gel electrophoresis or by transforming competent bacteria and plating in the presence of ampicillin, X-gal, and IPTG to assess the number of recombinant (white colonies) versus non-recombinant (blue colonies).

17. A ligation can be either stored frozen and aliquots provided to students or can be transformed into bacteria and miniscreen analysis used to obtain a small collection of recombinants of the desired sizes.

Appendix V

Commercial suppliers

This is not meant to be an all-inclusive list of commercial suppliers of the chemicals, reagents, and biologicals, but provides ordering information regarding suppliers with whom the author has experience obtaining the materials used in Chapter 12.

1. Bio-Rad Laboratories

USA
2000 Alfred Nobel Drive
Hercules, CA 94547-9980
USA
Telephone: (510) 741-1000
1-800-4-BIORAD
1-800-424-6723
FAX: (510) 741-5800
1-800-879-2289
Internet: http://www.bio-rad.com

UK/Europe
Bio-Rad House
Maylands Avenue
Hemel Hempstead
Hertfordshire HP2 7TD UK
Telephone: 01442-232552
Toll-free: 0800-181134
FAX: 01442-259118

2. Boehringer Mannheim Corporation

USA
9115 Hague Road
PO Box 50414
Indianapolis, IN 46250-0414
USA
Telephone: 1-800-262-1640
Voice mail: 1-800-262-1640
FAX: 1-800-428-2883
Internet: http://biochem.boehringer.com

UK/Europe
Boehringer Mannheim House
Bell Lane
Lewes
East Sussex BN7 1LG UK
Telephone: 01273-480444

3. Fisher Scientific

USA
711 Forbes Avenue
Pittsburgh, PA 15219-4785
USA
Telephone: 1-800-766-7000
FAX: 1-800-926-1166
Internet: http://www.fisher1.com

UK/Europe
Bishop Meadow Road
Loughborough
Leicestershire LE11 5RG UK
Telephone: 01509-231166
FAX: 01509-231893

4. GIBCO BRL/Life Technologies Inc.

USA
PO Box 6009
Gaithersburg, MD 20897-8406
USA
Telephone: 1-800-828-6686
FAX: 1-800-331-2286
Internet: http://www.lifetech.com

UK/Europe
3 Fountain Drive
Inchinnan Business Park
Paisley PA4 9RF UK
Telephone: 0141-8146100
FAX: 0141-8146317

5. New England Biolabs, Inc.

USA
32 Tozer Road
Beverly, MA 01915-5599
USA
Telephone: (508) 927-5054
Toll-free: 1-800-632-5227
FAX: (508) 921-1350
Internet: http://www.neb.com

UK/Europe
67 Knowl Piece
Wilbury Way
Hitchin SG4 0TY UK
Telephone: 01462-420616
Toll-free: 0800-318486
FAX: 01462-421057

6. Promega Corporation

USA
2800 Woods Hollow Road
Madison, WI 53711-5399
USA
Telephone: 1-800-356-9526
FAX: (608)-277-2516
Internet: http://www.promega.com

UK/Europe
Delta House, Enterprise Road
Chilworth Research Centre
Southampton SO1 7NS UK
Telephone: 01703-760225
FAX: 01703-767014

7. Sigma

USA
PO Box 14508
St. Louis, MO 63178
USA
Telephone: 1-800-325-3010
FAX: 1-800-325-5052
Internet: http://www.sigma.sial.com

UK/Europe
Fancy Road
Poole
Dorset BH12 4QH UK
Telephone: 01202-733114
FAX: 01202-715460

8. Stratagene

USA
11011 North Torrey Pines Road
La Jolla, CA 92037
USA
Telephone: 1-800-424-5444
FAX: (619)-535-5400
Internet: http://www.stratagene.com

UK/Europe
Cambridge Science Park
Milton Road
Cambridge CB4 4GF
Telephone: 01223-420955
FAX: 01223-420234

Index

D

G

H

I

K

L

R

U

V

W

X

Y

Z

Other Books of Interest

- **Molecular Biology: Current Innovations and Future Trends**

Editors: A. M. Griffin and H.G. Griffin
Part 1 1995, 165 pages. ISBN 1-898486-01-8 £19.99 (paperback). ISBN 1-898486-13-1 £49.99 (hardback)
Part 2 1995, 176 pages. ISBN 1-898486-03-4 £19.99 (paperback). ISBN 1-898486-14-X £49.99 (hardback).
A major two-volume work to keep you up-to-date with all the current technology in molecular biology. Each volume is packed with protocols and information on the most widely used techniques. In particular the book aims to keep you informed of the newest techniques and the most recent advances.

- **Internet for the Molecular Biologist**

Editors: S.R. Swindell, R.R. Miller, G.S.A. Myers
1996, 187 pages. ISBN 1-898486-02-6 £19.99 (paperback). ISBN 1-898486-10-7 £49.99 (hardback)
Written specifically for the molecular biologist, this indispensable practical guide aims to demystify the information superhighway by answering the questions: What is the Internet? What is it for? How can I do it? How useful is it for molecular biology? The book contains a wealth of information and practical instruction vital for the researcher of today. Even the experienced surfer will find an abundance of novel material in this invaluable guide. "...the Maniatis of the Internet..." *EmbNet News*
"I heartily recommend this volume for anyone who uses or wishes to use the Internet." *Heredity*.

- **The Lactic Acid Bacteria**

Editors: E.L. Foo, H.G. Griffin, R. Möllby, and C.-G. Hedén
1993, hardback, 102 pages, ISBN 1-898486-04-2 £49.99 (hardback).
The published papers of the first lactic acid bacteria computer conference. An essential reference text. This book represents a broad review of current lactic acid bacteria research with contributions from a number of well-known scientists.

- **Genetic Engineering with PCR**

Editor: Robert M. Horton and Robert C. Tait
September 1997. ISBN 1-898486-05-0 £34.99 (paperback). ISBN 1-898486-12-3 £59.99 (hardback).
While there have been myriad books written on site-directed mutagenesis, and portions of several PCR books address the topic, this is the first laboratory manual that focuses on synthetic (as opposed to analytical) uses of PCR. The entire volume is devoted to cover the topic in breadth and depth and includes a variety of approaches for mutagenesis, recombination, and the construction of synthetic or semi-synthetic genes.

- **Prions: Molecular and Cellular Biology**

Editor: David A. Harris
December 1997. ISBN 1-898486-07-7 £74.99 (hardback)
This major new work contains reviews of recent and current research on the genetics, cell biology, and biochemistry of prions and prion diseases. Comprehensive coverage. An essential text to all scientists working in this field. An outstanding resource!

- **Gene Cloning and Analysis: Current Innovations**

Editor: Brian C. Schaefer
1997, 214 pages. ISBN 1-898486-06-9 £ 34.99 (paperback). ISBN 1-898486-11-5 £ 59.99 (hardback)
This volume focuses on newly emerging technologies that facilitate the isolation and characterization of genes. The detailed protocols will be useful to the seasoned professional and easily understood by the novice. The vast majority of methods are applicable to any biological system, ranging from bacteria and fungi to plants, yeast, and higher eukaryotes. The protocols are accompanied by helpful notes and troubleshooting tips to ensure their successful execution. Additionally each chapter includes concise background information and frequently includes a review of similar methods followed by an assessment of future innovations.

Full details of all our books can be found at http://apollo.co.uk/a/horizon

Order Form

Horizon Scientific Press, PO Box 1, Wymondham, Norfolk, NR18 0EH, U.K.

No. of copies	ISBN	Title	Price
		Gene Cloning and Analysis: Current Innovations	£
		Internet for the Molecular Biologist	£
		Molecular Biology: Current Innovations and Future Trends Part 1	£
		Molecular Biology: Current Innovations and Future Trends Part 2	£
		Genetic Engineering with PCR	£
		Prions: Molecular and Cellular Biology	£
		The Lactic Acid Bacteria	£
		An Introduction to Molecular Biology	£
		Sub-total	£
		Add postage and handling £3 for one book, £5 for two or more books	£
		GRAND TOTAL	£

Method of Payment

☐ I enclose a cheque in UK£ drawn on a bank in the UK. (Eurocheques in UK£ also acceptable.) Please make all cheques payable to Horizon Scientific Press.

☐ Please debit my Mastercard/Visa/Eurocard
Card number:
Expiry date:
Your Name: (as on card)
Signature:

Shipping Information

Name:

Address:

Tel.

Email:

☐ **Mail your order to:**
Horizon Scientific Press
PO Box 1, Wymondham
Norfolk NR18 0EH
England

☐ **Fax your order to 01953-603068 (International +44-1953-603068)**

☐ **Email your order to: horizon@usa.net**

☐ **Order on-line at http://apollo.co.uk/a/horizon**

In the USA/North America order from: Horizon Scientific Press, c/o ISBS, 5804 NE Hassalo Street, Portland, Oregon 97213-3644.
Tel: (503) 287-3093 or (800) 944-6190. Fax: (503) 280-8832. Email: orders@isbs.com

In Australia/New Zealand order from: DA Direct, 648 Whitehorse Road, Mitcham, 3132 Australia.
Tel. (03) 9210-7777. Fax. (03) 9210-7788. Email: service@dadirect.com.au

All other countries order from: Horizon Scientific Press, PO Box 1, Wymondham, Norfolk, NR18 0EH, U.K.
Tel/Fax 01953-603068. International Tel/Fax +44-1953-603068. Email: horizon@usa.net